CHINA, OIL, AND ASIA:

Conflict Ahead?

CHINA, OIL, AND ASIA:

Conflict Ahead?

Selig S. Harrison

A study from The Carnegie Endowment
for International Peace

Columbia University Press · New York · 1977

Library of Congress Cataloging in Publication Data

Harrison, Selig S.
  China, oil, and Asia, conflict ahead?

  "A study from the Carnegie Endowment for
International Peace."
  Includes bibliographical references.
  1. Petroleum industry and trade—China.
2. China—Foreign relations—1949- I. Carnegie
Endowment for International Peace. II. Title.
HD9576.C52H37   338.2'7'2820951   77-8185
ISBN 0-231-04378-3
ISBN 0-231-04379-1 pbk.

Columbia University Press
New York   Guildford, Surrey

The Carnegie Endowment
for International Peace

The Carnegie Endowment for International Peace was established in 1910 in Washington, D.C. with a gift of $10 million from Andrew Carnegie, for the purpose of promoting international peace and understanding. In 1929 the Endowment was incorporated in New York.

As an operating (not a grant-making) foundation, the Endowment conducts its own programs of research, investigation, discussion, publication, education, and training in such international affairs fields as arms control, humanitarian policy, the Middle East, pre-crisis fact-finding, and international law and organization. The Endowment is engaged in several co-located joint ventures with other tax-exempt organizations to reinvigorate and extend the domestic and foreign dialogue on world affairs issues. The Endowment is associated with the publication of the quarterly *Foreign Policy*.

The Endowment's work is conducted from its centers in New York and Washington, and by its representative in Geneva, Switzerland.

77651

Contents

List of Illustrations

Foreword

As the worldwide search for new petroleum sources has intensified, one of its most significant but least understood consequences has been the spectacular increase in the importance of seabed, or "offshore," oil and gas development. More than 24 percent of the world's crude oil was produced offshore in 1976, and nearly one half of global undiscovered reserves are believed to be in undersea deposits. Already, twenty-eight countries have production under way offshore. Fourteen others have made discoveries that could lead to commercial production. Sixty more are actively searching for new offshore reserves. In the non-Communist world alone, oil companies and governments have announced plans to spend $106 billion on offshore exploration and development between 1975 and 1980.

This explosion of offshore activity has been an economic bonanza for many of the countries where oil has been found, but it has also created serious political problems, giving new sensitivity to long-simmering disputes over sea boundaries. The dangers latent in such disputes have been demonstrated on a limited scale in the 1976 Greek-Turkish conflict in the Aegean and in the reappearance of the Barents Sea controversy between Norway and the Soviet Union. Looking ahead, one can safely predict a proliferation of such conflicts over offshore resources. Some of the prime examples of these are likely to be the incipient disputes between China and its nine maritime neighbors.

In this revealing account, based on a two-year study for the International Fact-Finding Center of the Carnegie Endowment, Selig S. Harrison presents the first comprehensive report on Peking's carefully non-publicized offshore oil and gas program. Harrison shows why a growing Chinese offshore capability could foreshadow significant clashes of interest with neighboring countries, affecting, in particular, the future of Taiwan and South Korea; the Sino–Japanese–Soviet triangle; Sino–Vietnamese and Sino–Filipino relations; and the operations of American and other foreign oil companies with concessions in disputed areas. Powerful economic factors reinforce the political and strategic considerations that lie behind China's offshore ambitions. Increasingly, Harrison reports, Chinese leaders are turning to offshore development as one of the keys to the fulfillment of their energy production targets and thus to the achievement of rapid economic growth within their chosen framework of "self-reliance."

The author concludes that Peking has a "better than fifty-fifty chance" of reaching its goal of a 400-million-ton annual crude oil production level by 1990—comparable to the Saudi Arabian level in 1974. By contrast to export-oriented Saudi Arabia, however, Harrison emphasizes the many constraints that will make it difficult for the Chinese to keep up with their burgeoning domestic energy needs. Given these difficulties, he cautions that Peking is not likely to export much of its oil to other countries, with the notable exception of Japan.

On this critical point, Harrison amplifies in this book his much-discussed *Foreign Policy* article, "China: The Next Oil Giant," which called public attention in 1975 to China's potential as a major petroleum producer for the first time but downgraded the possibility that Peking would be a major exporter. Harrison's projections in mid-1975 were more conservative than those of the CIA and others at that time with respect to Chinese export levels, reflecting a cautious estimate of short-term production prospects. In projecting long-term production capabilities, however, his appraisal in these pages strikes a somewhat more bullish note than have recent CIA studies, while continuing to stress that Peking is unlikely to have large export surpluses. Indeed, his analysis suggests that China's current efforts to lay the foundations for energy self-sufficiency could pay off during the very period when the United States and the USSR begin to face the shortages predicted by President Carter.

Selig Harrison is one of the first Senior Associates to be appointed to the Carnegie Endowment's new International Fact-Finding Center in New York. He brings to this assignment a rich background in scholarly journalism with a broad stream of articles and books to his credit. His academic work has been done at Harvard, Columbia, and the Brookings Institution. As a journalist he has been a foreign correspondent in many Asian posts, has served as the Washington *Post* bureau chief in New Delhi and Tokyo, and is also a former managing editor of the *New Republic*.

In its methodology, this study exemplifies the research techniques which the Fact-Finding Center intends to use in studying emergent issues on the world scene. While extensive documentary material is cited, Harrison has relied primarily on interviews with 313 primary sources in Asia, Europe, and the United States. These sources have included officials of China and the other Asian countries directly concerned; United Nations experts; participants in the Law of the Sea treaty discussions; executives and geologists in American, Japanese, British, and other private oil and drilling companies; geophysicists, petroleum economists, and oceanographers; oil equipment firms that do business with Peking; and China-watchers in a variety of countries both in government and academic circles. Most of these sources have cooperated on the understanding that their anonymity would be preserved.

The International Fact-Finding Center consciously seeks to combine the investigative techniques of the responsible, serious journalist with the detachment and analytical discipline of the academic researcher. Founded in 1974, the Center attempts to anticipate "near-horizon" trends that could lead to international conflict or to large-scale human suffering. By presenting an independent, factual appraisal of such situations in their early stages, it hopes to focus attention by government and public-opinion leaders in the countries studied, as well as in the larger international community, while there is still time for ameliorative action.

In warning of possible trouble ahead, the Center is not necessarily predicting trouble. On the contrary, to the extent that it has the desired impact, a successful Fact-Finding study will confound its own projections. Here Harrison's investigation has also been a case in point, for his extensive interviews, together with the publication of his initial findings in *Foreign Policy*, touched sensitive nerves and brought into the open a problem that had hitherto been left to the private manipulations of the parties directly concerned. It can be argued that this public exposure has already helped to relieve tensions and to induce a mood of caution on the part of East Asian governments and the foreign oil companies involved. This book should further encourage negotiated settlements of the boundary disputes between China and its neighbors, opening the way for critically needed petroleum development in areas where drilling activity is now all but paralyzed by political confusion and uncertainty.

As always, Endowment sponsorship of the report implies a belief only in the importance of the subject. The views expressed are those of the author. Comments or inquiries on this and other work of the Endowment are welcome and may be addressed to the Carnegie Endowment for International Peace, 345 East 46th Street, New York, New York 10017 or 11 Dupont Circle, N.W., Washington, D.C. 20036.

Thomas L. Hughes
President
Carnegie Endowment
for International Peace

Acknowledgments

I should like to express my appreciation to the many individuals and organizations whose cooperation has made this work possible. Since more than three hundred persons were interviewed, many of them more than once and most of them on the understanding that their names would not be used, it is not possible to single out those who have been particularly helpful with their time and counsel. However, I should like to express appreciation to the Gulf Oil Company and the Amoco International Oil Company and their legal consultants, North-cutt Ely and Carl McFarland, for making available advisory legal opinions that have helped me to characterize the Law of the Sea assumptions underlying their concessions in Taiwan. Jerome A. Cohen, director of East Asian Legal Studies at the Harvard Law School, and Choon-ho Park, research associate in East Asian Legal Studies, were cooperative in the difficult early stages of my investigation. I am indebted to Dr. Park, to

J. Ray Pace, president of the Baker Trading Company of Houston, and to Thomas Franck, director of the International Law Program of the Carnegie Endowment, for their helpful comments on early drafts of the manuscript.

A special word of thanks is due to Frank and Clare Ford, of the Ford Studios, Arlington, Virginia, who were painstaking and patient in preparing eleven complex maps under my direction.

Sharon Rhoades of New York City did the translations of Japanese materials; Chinese translations were done by Richard Yiu of Hong Kong and Chiang-chung Chang and Lloyd Richardson of Washington.

I am grateful to many colleagues at the Carnegie Endowment for their generous support and encouragement, in particular Thomas L. Hughes, president, and Charles W. Maynes, former secretary. I relied heavily on Diane Bendahmane, assistant director of publications, who supervised the editorial work; Margaret Ameer Cataldo, administrative assistant at the International Fact-Finding Center, who performed myriad research tasks; and Susan Fisher, who typed the manuscript and made many helpful editorial suggestions. The Columbia University Press was cooperative beyond the call of duty in processing the book and expediting its publication.

S. S. H.
June 1977

CHINA, OIL, AND ASIA:

Conflict Ahead?

CHAPTER ONE

The *Gulfrex* Decision

It was New Year's Eve, and the eighteen American officials who had hurriedly gathered in the State Department Operations Room on 31 December 1970, were still sharply divided after four hours of discussion.[1] The issue was what, if anything, the Seventh Fleet should do in the event of Chinese naval action against the U.S. seismic survey vessels then beginning to explore for oil in disputed waters of the East China Sea, the Taiwan Strait, and the Yellow Sea. For more than a year, the State Department and the White House had watched nervously as South Korea and Taiwan had allocated concessions to American and other foreign oil companies in offshore areas claimed by Peking or Tokyo or both. Their anxieties had grown in mid-November when a sophisticated Gulf survey ship, equipped with top-secret electronic and inertial navigation equipment, had started to explore north of Taiwan in an area only forty to fifty miles off the coast of

China. Peking had looked the other way, but the latest word was that Komar-class Chinese patrol boats, armed with Styx missiles, were closely shadowing the *Gulfrex* and another nearby vessel that was also equipped with sensitive apparatus. One Chinese boat was never more than half a mile away from the Gulf ship. Fearful of a new *Pueblo*-type incident, the American Embassy in Taipei and CINCPAC (Commander-in-Chief, Pacific) headquarters in Honolulu were urgently pressing for instructions.[2] Peking was also attempting to send diplomatic warning signals to Washington, and a disturbing intelligence report had just been received shortly before the meeting convened. In a dinner-party conversation with the foreign minister of a third country, a senior Chinese ambassador had gone out of his way several times to emphasize that Peking meant exactly what its increasingly strident propaganda pronouncements had been saying on the subject of offshore oil rights. China regarded the entire East Asian continental shelf as an extension of Chinese territory and took a "grave view," accordingly, of "American oil imperialism in collusion with Taiwan."

To representatives of the Joint Chiefs of Staff and the three services, it was self-evident that the United States should respond with the appropriate level of nonnuclear military force if China attacked or harassed U.S. survey vessels. The credibility of the American deterrent would be destroyed throughout the Far East if China could act with impunity. At one pole of the discussion, Rear Admiral William R. Flanagan assumed a combative, uncompromising posture that reflected the tensions then developing between the Pentagon and the White House in the prelude to the Nixon reversal of U.S. China policy. In late 1970, it will be recalled, the United States had initiated its first modest steps to open trade with the Communist regime, foreshadowing the secret Kissinger visit to Peking in mid-1971. The next gesture to the Chinese was generally expected to come on the military front. By taking a strong stand on the issue of survey ships, Admiral Flanagan was responding, indirectly, to White House suggestions that the Seventh Fleet patrol in the Taiwan Strait might be thinned out or even eliminated. The defense of Taiwan and its offshore oil reserves was not only important in regional military terms, the admiral argued, but should be a key element in a larger political-military strategy directed to reducing American dependence on the Middle East. The United States had the most advanced technology in the world for finding offshore oil, he contended, and should make the most of this advantage. By the same token, much of the sensitive equipment used for oil exploration had direct military applications and could be kept from hostile hands only by close teamwork between private survey vessels and supporting naval craft. At the insistent request of Gulf's representatives in Washington and Taipei, who felt strongly with regard to the U.S. government's obligations in the matter, arrangements for such teamwork had already been made in the case of the *Gulfrex,* and the ship deserved full protection, he argued, unless the United States was prepared to abandon Taiwan to the Communists.

Nominally, Admiral Flanagan was the ranking spokesman at the meeting for the Defense Department as a whole, representing the Bureau of International Security Affairs in the absence of his civilian superior, Assistant Secretary Warren Nutter. As the debate grew more and more tangled, however, it soon became apparent that the Pentagon did not have a unified position. The meeting had been called on short notice at a time when all of its top officials were away for the holidays. With Secretary of Defense Melvin Laird out of the country and Deputy Secretary David Packard visiting on the West Coast, authority had fallen by default to Nutter and his aides, who occupied the far Right on the Pentagon ideological spectrum and frequently lost out in internal departmental policy controversies. A militant anti-Communist, the controversial assistant secretary was known, in particular, for his advocacy of a more assertive U.S. military role in Asia. Admiral Flanagan found himself openly challenged not only by State Department critics but also by Pentagon civilian officials, notably Leigh Ratiner, then a young legal adviser specializing in Law of the Sea issues. Defense and State Department lawyers both stressed the delicacy of the interlocking boundary disputes in the East China Sea and adjacent waters between China, Taiwan, Japan, and South Korea. At the same time, the State Department was itself internally divided, with some of the strongest support for Flanagan coming from U. Alexis Johnson, undersecretary of state for political affairs, and much of the most determined opposition coming from economic officials. Assistant Secretary for Economic Affairs Philip H. Trezise minimized the importance of a much-discussed United Nations seismic survey report pointing to the "high probability . . . that the continental shelf between Taiwan and Japan may be one of the most prolific oil and gas reservoirs in the world." [3] Even if this proved to be true, he declared, it would be many years before definitive geological judgments could be made on the basis of actual drilling, and the Middle East would continue to provide the overwhelming bulk of U.S. oil imports for an indefinite period.

Supporting the Trezise estimate, Robert Hormats, representing the National Security Council, observed that the uncertain economic stakes involved would justify, at most, a carefully circumscribed military response, especially in view of the obvious political risks. But the White House, too, had been caught unprepared, and Hormats failed to give a firm lead. With nerves frayed and New Year's Eve parties beckoning, the stalemated meeting appeared ready to accept the idea of a "temporary" compromise that would have permitted local commanders to use nonnuclear force in the event that the Chinese actually attempted to board and seize a survey vessel. No action would have been authorized in cases of low-level harassment, and another meeting was to be held within ten days. At that point, to the surprise of all concerned, Ratiner suddenly asked for a recess and persuaded Flanagan to join him in submitting the issue to Deputy Secretary Packard, then officiating in the absence of Secretary Laird overseas.

What followed set the pattern for a hands-off U.S. military policy toward a variety of offshore oil conflicts in East Asia that were to grow in·sensitivity during the uneasy years of transition to a new American relationship with China. Packard was at his residence in Palo Alto, California, dressing to go out for the evening, when the call came from Flanagan and Ratiner. Unbeknown to the admiral, Packard had been briefed several hours earlier and knew that the Washington meeting was taking place; and unbeknown to Ratiner, who was not sure what to expect, he had made an unequivocal decision against military intervention of any kind. As Packard recalled it, he was "not aware, at that time, of the contemplated move by President Nixon with respect to the establishment of contacts with Peking." Rather, it was simply a matter of keeping out of jurisdictional conflicts, and the Gulf vessel had, after all, "been warned in advance that it was operating in disputed waters. I concluded that was the extent of our responsibility, and under such circumstances it would be improper to ask our Navy to provide protection." [4] In addition to overruling Flanagan on the immediate issue of intervention, Packard also added his personal judgment that all American oil companies active in the area should be specifically advised to stay out of disputed East Asian waters in the future. This was enough to tip the scales when the meeting resumed, and by 10:35 P.M. an agreed cable to Honolulu had been drafted ruling out the use of U.S. forces in support of Gulf's exploration activity. [5] At the same time the State Department was instructed to inform oil-company representatives in Washington and U.S. embassies throughout Asia that the American posture in disputes relating to oil survey operations would thereafter not only be one of scrupulous noninvolvement but of active discouragement. If American companies nonetheless conducted survey operations in disputed areas, they would do so at their own risk. They were not to use U.S.-flag survey vessels or drilling rigs; not to employ U.S. citizens in the crews of such vessels; not to use classified U.S. technical equipment that might fall into the hands of other military forces; and not to expect cooperation in using U.S. satellites for navigation purposes.

The significance of the Packard decision was underlined by the fact that the Joint Chiefs of Staff (JCS) seriously attempted to reverse it for three months thereafter, insisting on the creation of a State-Defense task force to study the matter further. In early January, Admiral Flanagan served formal notice in messages to the State Department that the JCS would continue to treat the 31 December decision as "interim guidance" to CINCPAC "prior to the return to the survey area" of those exploration vessels that had temporarily returned to port in Taiwan.[6] It was only after a White House-appointed Senior Review Group had backed it up that the Packard decision was publicly unveiled as U.S. policy in a formal State Department announcement on 9 April 1971.

In retrospect, this was a critical link in the chain of minidecisions that set the stage for an overall shift in China policy, a chain that also included the adoption of a neutral stand in the territorial dispute over the coveted Senkaku Islands (known as Tiao-yü T'ai in Chinese) between China, Japan, and Taiwan. Having acquired the Senkakus (Tiao-yü T'ai) from Japan along with the Ryukyu Islands under the San Francisco peace treaty, the United States felt compelled to return them to Tokyo as part of the then pending Okinawa reversion agreement, especially in the wake of seismic surveys pointing to rich undersea oil reserves in the area. To the dismay of Japan, however, the United States also made clear that it recognized the existence of a controversy over the title to the islands and did not mean to indicate support for the Japanese claim by including them in the reversion package. The attempt to avoid embroilment in the Senkaku (Tiao-yü T'ai) dispute went hand in hand with the Packard policy of military detachment in oil survey operations, and the disclosure of the New Year's Eve decision was coupled with a formal declaration of neutrality in the Senkaku (Tiao-yü T'ai) issue.[7] In both cases, China policy was the underlying issue but had not yet been explicitly defined as such in the confused policy jockeying that preceded the August breakthrough. The 9 April 1971 announcement was followed soon afterward by Peking's "ping-pong" initiative and then by the carefully orchestrated reduction of U.S. naval patrols in the Taiwan Strait on the eve of the Kissinger visit.

Was there ever a serious danger of a *"Gulfrex* incident" in the East China Sea? Gulf sources stress that the ship never went remotely close to Chinese territorial waters and was promptly called back to port in Taiwan following the Packard decision. If exploration had continued, however, and if U.S. naval vessels had been deployed protectively nearby, it is difficult to say what might have happened and how the 1971 opening to China might have been affected. Peking's wide-ranging offshore resource claims are not limited to its territorial waters, it should be kept in mind, or even to the 200-mile zone envisaged in the projected Law of the Sea treaty, but explicitly embrace the entire breadth of the continental shelf as well as other areas defined by historic claims to islands and coral reefs (Figure 1). Peking contends that the sedimentary deposits on the shelf originally came from Chinese rivers washing into the sea and are thus rightfully Chinese.

As this book seeks to demonstrate, subsequent developments have sharply underlined the potential importance of offshore oil for China and the increasing sensitivity of Peking's offshore disputes with neighboring countries. The much-publicized Chinese takeover of the Paracel Islands is only one of four cases to be discussed here in which oil-related offshore disputes over territorial or sea boundary issues have provoked the use or implicit threat of force not only by Peking but also by Hanoi, Pyongyang, and even Tokyo. Gulf, for example, disregarding State Department warnings, decided to take its chances and conduct survey and drilling operations in a South Korean concession in the Yellow Sea during late 1972 and early 1973. North Korean "fishing" boats responded by cutting the company's seismic cables repeatedly, and Chinese patrol boats came within a mile of a Gulf drilling rig on three separate occasions in a rerun of the *Gulfrex* incident that once again led to intervention by Washington.

Despite the relaxation of tensions in East Asia following the establishment of Sino-U.S. contacts, the indeterminate status of offshore petroleum rights in the East China Sea, the Yellow Sea, the Taiwan Strait, and the South China Sea has continued to complicate relations between Peking and the nine neighboring states that also have claims of varying magnitude to the resources beneath these waters.[8] Indeed, the dramatic changes that have been taking place in the global economic environment have made offshore resource disputes much more serious in character than they were in 1971. As countries in the area have become more and more aware of their dependence on Middle East producers, so their hunger for domestically controlled petroleum sources has intensified. For Taiwan, South Korea, and the former Saigon regime, in particular, offshore oil development has been viewed as a key to economic survival. For Japan, too, the rise of the Organization of Petroleum Exporting Countries (OPEC) has been a traumatic blow. Tokyo is increasingly torn between the urge to develop the continental shelf independently, even at the risk of conflict with Peking, and a desire for a new energy partnership with China that would entail acquiescence in Peking's shelf claims as the quid pro quo for assured oil imports. Moreover, as the probable dimensions of the offshore potential in the area have gradually been confirmed by geological findings, the appetites of all concerned have grown. China has been stepping up its preparations for offshore activity and will soon have the capability to survey and drill in most of the areas contested by its neighbors. In an ironic footnote to the events of December 1970, Peking even made an unsuccessful attempt in 1975 to buy the *Gulfrex* for its offshore program.

As for the American oil companies with involvements in the region, the moderation of tensions between Peking and Washington has not deterred some of them from exploration plans in disputed areas, to be examined in later chapters. On the contrary, the Peking-Washington dialogue has aroused hopes among these companies that the United States will be able to use oil as a bargaining counter. In return for the "derecognition" of Taiwan and for helping China to achieve a balance of power vis-à-vis the Soviet Union, it has been argued, Washington should be able to extract a double-edged oil bargain from Peking: on the one hand, American companies should be able to obtain crude oil from Peking in payment for technical assistance in developing uncontested offshore areas close to the mainland; on the other, they should have freedom to explore in disputed areas under the aegis of South Korea and Taiwan. Peking, with its "self-reliance" commitment, has shown little interest in such direct and explicit arrangements with either the Western majors or the Japanese oil companies that also have their eye on the continental shelf. But there have nevertheless been hints that China may be prepared for sea boundary adjustments, and the resolution of offshore resource controversies has become increasingly intertwined with the larger search for security and political accommodations in East Asia.

This book begins with a discussion of China's energy potential, focusing on the importance of offshore oil in China's plans, followed by detailed appraisals of the offshore reserve prospects in areas claimed by China and of Peking's emerging offshore capabilities. Chapters 5 to 7 examine the offshore ambitions of Taiwan, South Korea, and Japan, assessing the potential for conflict with China posed by their offshore development plans. The South China Sea is treated as a special case in which strategic factors and historically rooted territorial disputes overshadow the competition for oil resources as such. Finally, chapters 9 to 11 consider East Asian offshore resource rivalries in relation to Law of the Sea issues as well as the policy choices open to China and its neighbors, to non-Asian major powers, and to the broader international community.

CHAPTER TWO

China's Oil Potential: Problems and Prospects

The potential for conflict over the oil and gas reserves of the offshore areas adjacent to China can be fully appreciated only in the context of the burgeoning energy needs generated by Peking's ambitious drive for economic development. It is the rapid multiplication of economic pressures in a country so vast and so populous that has prompted Peking to augment its onshore search for petroleum with an ever-expanding offshore exploration program as well. As we see later, strategic and political considerations have accelerated the Chinese decision to "go offshore," and Peking is actively preparing to explore in disputed waters as a means of forestalling foreign incursions and staking its own claims. Given the dimensions of the Chinese development challenge, however, Peking also has a profound, long-term economic stake in how the resources of the continental shelf are ultimately divided. This chapter examines the extent of emerging Chinese energy needs, the reasons for the shift from coal to oil as China's primary energy source, and the critical importance of the off-shore factor in assessing how rapidly China will be able to expand its petroleum production.

Oil and Coal
in a Changing Economy

The probable magnitude of future Chinese energy needs has been suggested by a variety of studies based on differing assumptions regarding future energy-consumption patterns that are governed, in turn, by differing analyses of past consumption as well as by a host of variables such as alternate possible economic growth rates and development strategies. Each of these projections must be assessed in relation to the particular set of assumptions utilized, but all of them underline the fact that a significant acceleration of the economic growth rate would mean massive increases in the demand for energy.

Tatsu Kambara, of the Japan Petroleum Development Corporation, who assumed an average annual economic growth rate of only 5 percent, slightly less than the 5.6 percent average recorded during the 1953–74 period, estimated that total energy demand would jump from the presumed 1974 level of 253.2 million tons to 422.7 million tons by 1982.[1] Another Japanese analyst, Masanobu Otsuka, of Tokyo's Nomura Research Institute, posited an 8.3 percent overall growth rate and a 10.1 percent industrial growth rate, slightly higher than the 9 percent average industrial rate recorded between 1958 and 1974. Otsuka projected an 807.6-million-ton energy demand by 1985.[2] Thomas Rawski, of the University of Toronto, using an 8.8 percent overall growth rate and a 10 percent industrial growth rate, projected 929 million tons by 1985.[3]

What these projections would mean in terms of petroleum depends on the emphasis given to the development of oil and gas relative to other energy sources by Chinese planners. Recent estimates indicate that oil has been used to satisfy 18 to 22 percent of Chinese energy needs, with coal believed to account for 68 percent at most, and hydroelectric power less than 2 percent. It will be my purpose here to show that China is moving increasingly from a coal-based to a petroleum-based economy and that the share of oil in Chinese energy usage is likely to double within the next decade. Accordingly, if one were to adopt the 48.2 percent estimate for oil use projected for 1985 by the Japan External Trade Organization,[4] Kambara's estimate for overall energy demand would mean a need for 204 million tons (1.5 billion barrels) of oil by 1982,[5] Otsuka's would require 389 million tons (2.9 billion barrels) by 1985, and Rawski's would imply 447 million tons (3.4 billion barrels) by 1985. The immensity of the challenge represented by these projections is apparent in the fact that Chinese oil production in 1976 totaled only 85 million tons (637 million barrels). Nevertheless, Peking has boldly embarked on a crash program to push this level upward as rapidly as it can, along with increases in natural gas production, and has correspondingly started to downgrade the relative emphasis given to its investments in other spheres, including the development of coal and other alternative sources of energy. The unannounced Chinese target for oil production is 400 million tons (3 billion barrels) by 1990,[6] which could permit a stepped-up economic growth rate, but one falling considerably short of 8 percent.

There are two principal reasons for the gradual shift that has been taking place in China from a coal-based to a petroleum-based economy. The first is the disproportionately higher cost of developing and utilizing coal, given the depleted condition of most existing mines and the location of many untapped deposits. Related to this factor is the changing character of the energy needs resulting from new approaches to economic development, not only in industry but in agriculture as well.

Chinese coal reserves are believed to be enormous, totaling an estimated 11 trillion tons, an amount equal to one-third of the known world reserves. With only half of the known veins so far exploited, it has long been assumed that a logical course for Chinese energy development would be to emphasize coal mining, especially since an estimated 71 percent of these already discovered deposits are located in parts of the northeast, north, and northwest, where Chinese industry is growing most rapidly (Figure 2). As mining activity has proceeded, however, geologists have discovered that the most accessible parts of the known veins in the northeast and adjacent industrial areas are either already depleted or will soon be depleted. Attempting to keep pace with the expanding needs of the northeast industrial belt by stepping up coal production there would entail ever more costly efforts to get at the less accessible seams of existing mines or to develop entirely new mines. An even more expensive alternative would be to develop new mines elsewhere, transporting the coal to other areas over an already taxed rail system.

Continued emphasis on coal as its primary source of energy has become less and less attractive as China has become more and more confident of its oil potential during the period of intensified exploration and production in the northeast and elsewhere that began in 1960.[7] As chapter 2 will show, Peking has discovered that in petroleum, too, its natural endowments are enormous. While new fields were opening up one after another in the northeast (Figure 2) and oil production was growing nationally at an average annual rate of 24.6 percent,[8] the annual rate of increase in coal production was gradually dropping from 8 percent in 1971 to 2.9 percent in 1974 and an estimated 3.5 percent in 1975.[9] In 1976 coal production started to regain some momentum, but the Tangshan earthquake kept the rate of increase down once again. From 1958 to 1965 oil accounted for only 12 percent of the growth in energy use as against 76 percent for coal; between 1971 and 1974, the figures were 41 percent for oil and 48 percent for coal.[10] Reporting on a major national mining conference in late 1975 the New China News Agency pointed to the "complicated natural conditions, geological variations and many dangers" in mining and concluded that "it is not easy to increase coal production quickly on a large scale." [11] The conference mapped an intensive ten-year program designed to mechanize existing mines and increase productivity but did not envisage the construction of major new mines on a nationwide basis. No overall growth targets were cited, and the emphasis was on developing small, local mines to serve new industries in the south and other less industrialized areas.

It is often suggested that hydroelectric and nuclear power might offer alternatives to oil and gas for industrial uses if cost factors should rule out a continued emphasis on coal. Initially, China under Communist rule did place great emphasis on the development of hydroelectric power until interference with water transport and the technical problems associated with storage led to second thoughts. Hydroelectric power is now expected to play a significant role in rural areas and in small-scale power generation but is not expected to serve as a substitute for coal in most industrial activity. Similarly, nuclear power has thus far been ruled out as uneconomic except, possibly, for a few urban centers where the demand for electricity proves great enough to permit the operation of large-scale units on a continuous basis.

Most evidence indicates that China is continuing to rely on coal for its thermal power stations in areas where coal is readily available and oil is not, but that there has been a gradual shift to oil and gas in other areas where a choice exists, as in the northeast industrial belt. Significantly, as a Soviet observer has noted, some 70 percent of all electric power in China is produced in the northeast, north, and east. [12] The fact that China burns some of its crude in power plants is one of the key reasons accounting for the seeming discrepancy between annual crude oil estimates and the amount of oil exported and consumed in refineries and petrochemical plants. [13] To some Western observers, burning unrefined crude as fuel is a wasteful practice. But others point out that coal has the lowest thermal efficiency among the major energy sources,[14] and that there are technical reasons, in any case, for this use of some of the particular grades of crude found in the northeast, given the present state of Chinese refining technology.[15]

Oil constituted only 5 percent of the fuel consumed by industry in 1957, but this figure jumped to an estimated 22 percent by 1974.[16] In addition to using crude oil as a fuel for power plants, China is consciously seeking to take advantage of its abundance of petroleum and petroleum-derived raw materials by building petrochemical and fertilizer plants. More than $1.4 billion worth of Japanese and Western equipment and, in some cases, complete plants has been imported for the expansion of the petrochemical industry since 1963, most of it following the resumption of contacts with the United States and Japan. Fifteen ammonia and urea complexes ordered since 1972 will draw on oil-related gas deposits and natural gas, primarily from Szechwan Province, marking an abrupt shift from what has hitherto been reliance on coal-based fertilizer manufacturing processes.

It should be noted parenthetically here that information relating to targets and production achievements in natural gas and shale oil is extremely fragmentary and unreliable. For this reason, no attempt has been made to deal directly with these areas in this book. Signs of growing interest in gas production have been noted, however, especially in connection with the fertilizer and petrochemical industries, and significant gas discoveries could conceivably lead to a marginal reduction in the 400-million-ton (3-billion-barrel) crude-oil target in order to release investment funds for gas.

The most dramatic of the new petrochemical complexes springing up throughout China are a $450-million installation at Liao-yang in Manchuria, which will draw on oil from the Taching field, and a mammoth grouping of six plants near Shanghai. Other major complexes already built or nearing completion in 1975 were tied in with refineries at Peking and

Nanking and with oil fields at Taching, Tientsin, Takang, Canton, and Maoming. Already, China is producing substantial amounts of polyvinyl chloride, polystryrene, and polymethyl and is beginning to make vinylon and polyethylene in limited quantities. A wide range of products from plastics and synthetic textiles to foam rubber and kitchen utensils is pouring out of these new petrochemical installations, many of them destined for export as well as for the domestic market. As the more recent equipment purchases are put to use, China will be able to produce 3 billion pounds of ethylene per year, foreshadowing a major expansion of its polyester fiber capacity.[17] The growing importance of the petrochemical industry in China's increasingly oil-centered energy policy was reflected in a major governmental reorganization in 1975 in which all aspects of petroleum development were put under the control of a newly expanded oil ministry after a long bureaucratic struggle between oil and coal officials. Pointing to the supremacy of oil officials in the revamped structure of the Ministry of Fuel and Petrochemicals (which embraces oil, coal, and other subdivisions), *Nihon Keizai*, the leading Japanese business daily, concluded that Peking had reached "a major economic turning point in its economic strategy, with coal giving way to oil in emphasis and petrochemicals assigned to serve as the vanguard of industrialization."[18]

As in the case of industry, the changing character of energy needs is vividly exemplified in the countryside, where Peking is putting intense emphasis on agricultural mechanization. A major national conference at Tachai in late 1975 called on all rural communes to accelerate the mechanization of farm work in order to step up production and maximize local self-sufficiency. Chinese statistics indicate a 20 percent yearly rate of growth in the amount of oil used for tractors and a 30 percent yearly growth in the amount used for irrigation pumps. Visiting farm communes or the Shanghai Industrial Exhibition, one finds that Chinese factories are turning out a bewildering variety of farm machinery, implements, and vehicles using gasoline, including a growing number of bulldozers, trucks, and jeeps. Truck manufacture rose to 100,000 units in 1973 and is increasing at a rate variously estimated to be between 6 and 10 percent.

The drive for farm mechanization is closely linked with an expansion of the rural transportation infrastructure that will knit together hitherto isolated farm areas for economic development purposes while making it easier, at the same time, to control the countryside politically. Peking announced a 50 percent increase in road mileage during 1974, including the completion of nine new truck roads, 13,000 miles of feeder roads in rural areas, and 6,200 miles of short access roads linking communes with feeder roads. This continuing road-building effort foreshadows not only the steadily increasing use of gasoline in motor transportation during the decades ahead but also the growing use of oil for asphalt topping. Beyond this, stepped-up efforts to extend railroad lines into the countryside have led to a multiplying demand for diesel oil, as trains shift from coal-powered to diesel-powered locomotives.

The Investment
Question Mark

Evidence abounds that China is relying increasingly on petroleum to meet its expanding energy needs. But there is room for debate regarding whether Peking will be able to achieve its target of a 400-million-ton (3-billion-barrel) annual production level by 1990. This outcome will hinge on several critical economic and political variables, notably, the degree of emphasis that China gives to oil from year to year in its budgetary allocations, the rivalry between "radicals" and "moderates," especially as it affects the import of foreign technology, and the rate at which new recoverable reserves are found, which will significantly depend, in turn, on how rapidly Peking pursues its nascent offshore exploration program.

For more than eleven years, Chinese crude production increased at an average annual rate of 24.6 percent, rising from 6.4 million tons (48 million barrels) in 1963 to 20 million tons (150 million barrels) in 1970 and an estimated 65 million tons (487.5 million barrels) in 1974. Production continued to increase in 1975 at a rate of 16 percent, reaching a level of 75 million tons (562 million barrels) for that year, but dropped to a 12.7 percent rate of increase during the first quarter of 1976 over the first quarter of 1975 and a 13 percent rate for the entire year. The drop in 1975–76 reflected in part a deliberate shift in water injection methods at Taching, to be discussed later in this chapter. Designed to avoid wastage, the new methods entail a slower rate of production in the areas where they are applied but could ultimately double the amount of oil obtained from those areas and would not necessarily slow down the overall rate of national production if new reserves are discovered rapidly enough. According to unconfirmed reports, the Tangshan earthquake may also have damaged casing in the wells at Takang. In general, official announcements implied that the 1975 slowdown was intentional and temporary, stressing that crude oil production *capacity* went up by 20 percent during the first half of 1976 over the comparable period of 1975 and that natural gas production capacity had gone up by 200 percent.

To attempt specific projections of Chinese production levels is a risky game fraught with more than the customary quota of the booby traps that bedevil petroleum futurology. Japanese spokesmen have often tended to inflate their estimates to improve their bargaining posture with OPEC and the Western majors, and the Chinese have encouraged Japanese optimism as part of their anti-Soviet strategy. At the other extreme, the Western oil companies, given their interests elsewhere, have generally minimized Chinese prospects and technical capabilities. However, even if the annual rate of increase were to remain as low as 12.7 percent, the resulting production level of 426.6 million tons (3.2 billion barrels) in 1989 would assure realization of the Chinese target.

The most euphoric predictions concerning Chinese prospects have been made by Ryutaro Hasegawa, chairman of the Japan-China Oil Import Council, who declared on his return from a purchasing mission to China in 1974 that Peking would produce 400 million tons (3 billion barrels) by 1980. By 1975 a more careful appraisal by the Japan External Trade Organization led to the projection of a 440-million-ton (3.3-billion-barrel) level by 1985, with the growth rate expected to drop from 23 percent in the 1975–78 period to 20 percent for the two years thereafter, and finally to 16 percent from 1981 to 1985.[19] The Ministry of International Trade and Industry in Tokyo suggested in an unpublished estimate that the rate of growth would gradually drop off to 11 percent after 1978 but that production would still reach a 450-million-ton (3.4-billion-barrel) level by 1988. Similarly, U.S. government projections have gradually been downgraded from a 226-million-ton (1.7-billion-barrel) estimate for 1980 [20] by a CIA analyst in May 1975, based on an assumed 23 percent growth rate, to an estimate of 15 percent and 160 million tons (1.2 billion barrels) in a CIA study released six months later.[21] This revised estimate was roughly parallel to an unpublished British government estimate and a projection by a U.S. Commerce Department consultant.[22] Looking beyond 1980, the CIA envisaged the possibility of a 10 percent average annual growth rate (i.e., 415 million tons, or 3.1 billion barrels, by 1990, assuming a 15 percent rate from 1975 to 1980) but observed that this would be a "remarkable achievement" in the light of competing demands on Chinese investment resources and growth rates in petroleum production in other countries.[23] The Commerce Department adviser was more confident about China's prospects, predicting a 335-million-ton (2.5-billion-barrel) level by 1985 and, by implication, well over 400 million tons (3 billion barrels) by 1988, a prediction similar to one by geologist A. A. Meyerhoff in 1977.[24]

My own analysis suggests that there is a better than

fifty-fifty chance for a 400-million-ton-plus (3-billion-barrel-plus) production level by 1990, barring a major reversal of the oil development policies evolving in 1976 and early 1977. Such an achievement would indeed be a remarkable example of what a totalitarian state can accomplish by pushing some sectors of development and starving others. Chinese production in 1990 would then be on a par with Saudi Arabian production [25] in 1974 (412 million tons, or 3.1 billion barrels). At the same time it should be emphasized that populous China, with its vast domestic demands, cannot become "another Saudi Arabia" in terms of its export capabilities and that it is not yet clear how Chinese reserves compare with those of the Middle East and the Persian Gulf.

The central factor affecting the rate of Chinese oil development will be the budgetary emphasis given to oil relative to other energy sources and competing nonenergy sectors of economic development. Since Peking has not published detailed economic statistics since 1960, hard information relating to budgetary planning is not available to support either optimistic or pessimistic projections. In general, however, it appears that a much higher priority is being accorded to investment in oil than to investment in coal. This impression was strongly reinforced by Chairman Hua Kuo-feng's resounding dictum on 4 May 1977 that "some 10 more oilfields as big as Taching" should be built by the turn of the century in order to make China's oil industry "surpass that of the United States in the not too distant future." The visitor to China soon senses that the atmosphere surrounding petroleum development is one of an all-out "Manhattan Project" in which the dominant theme is the rapid expansion of an ambitious physical infrastructure. Numerous Japanese and other foreign businessmen who have met with Chinese oil officials and visited oil fields and oil-related facilities report evidence of crash programs to train technicians, build factories for the indigenous manufacture of petroleum equipment, find new reserves, open up new fields, convert power plants from coal to oil, and build refineries and petrochemical complexes fast enough to handle their share of the increase in oil production. By contrast, as noted earlier, the coal industry is at a stage of consolidation, with efforts to step up output concentrated largely in existing mines.

The coal-to-oil shift now taking place in China logically suggests that capital allocations for oil development are rising relative to coal and that Peking anticipates a corresponding leveling off of the rate of increase in coal production in order to release funds for oil. Accordingly, the Japan External Trade Organization assumed only a 5 percent growth rate in coal in making its projections of high growth rates in oil.[26] In the light of 1974 and 1975 data on coal production, it even appears possible that Peking has made a conscious decision to hold coal to a 4 percent growth rate. Why, then, have some analysts built their assessments of Chinese oil prospects on the assumption that coal growth targets would have to be set at present or upgraded levels? For example, reservations expressed by the CIA with respect to growth after 1980 were based on this assumption. The "median" CIA projection for 1980, cited earlier, juxtaposed a 15 percent oil growth rate with a 6.5 percent coal growth rate. Alternative CIA scenarios illustrated a less likely "high" growth rate by bracketing 20 percent for oil with 7 percent for coal and a "low" growth rate by combining 10 percent for oil with 6 percent for coal. Having assumed these high growth rates for coal, the CIA study then reached its conclusion that the competition for funds between coal, oil, and such bottleneck industries as steel would permit only "constant or slightly growing investment in the petroleum industry." [27] But why not 10 percent or 15 percent for oil juxtaposed with, say, 4 percent for coal?

The answer to this question depends upon the fundamental assumptions made regarding the basic thrust of Chinese development strategy and where oil fits into this strategy. Many foreign observers take it for granted that a Chinese attempt to expand oil production rapidly would be motivated by a desire to earn foreign exchange through exports in order to finance the import of needed industrial technology. In this perspective it would indeed be necessary to step up coal production as rapidly as possible to meet a major portion of domestic energy needs, while simultaneously accelerating oil production in the hope of securing export surpluses. Thus, addressed as it was to the issue of Peking's export potential, the CIA study assumed high growth rates for coal and oil alike, concluding quite plausibly that China would not be able to export oil on a scale comparable to the OPEC countries. There would simply not be enough capital available to generate massive export surpluses, the study argued, unless Peking was prepared to make drastic sacrifices in its short-term economic growth or to accept some form of foreign investment. To focus on Peking's export potential, however, is to lose sight of the emphasis on "self-reliance" that still underlies the Chinese approach to development. If one assumes, instead, that Peking intends to use most of its oil domestically, it is plausible to anticipate rising capital allocations for oil relative not only to coal but also to selected nonenergy sectors of development. In these nonenergy sectors, the short-term damage to growth resulting from reduced investment inputs would be less drastic than the damage that would result from an export-oriented energy policy and would be offset, in any case, by the economic advantages of oil as a more efficient and more versatile fuel than coal.[28]

It would be beyond the scope of this study to attempt an assessment of the merits and demerits of "self-reliance" compared to other approaches to development. What should be taken into account here is simply that this approach is a fact of life in the case of China, an authentic, broad-based expression of Chinese nationalism and not, as it is often pictured, merely the ideological aberration of a narrow, factional group. In a nationalist perspective, high growth rates and rising production levels are valued primarily in relation to the larger political objective of building up China as an independent power factor in the global arena. It is better to go more slowly and keep control of what is developed economically than to achieve a rapidly rising gross national product at the cost of foreign dependence. To be sure, there are significant differences between "radicals" and "moderates" with respect to the implementation of the "self-reliance" policy, and the downfall of the "Gang of Four" has greatly strengthened technocrats who are prepared to depart from the policy for the sake of specific technological, financial, or diplomatic purposes. In particular, the need for certain critical technological imports is acknowledged even by the most determined advocates of autonomy, which provides a rationale for exports of crude oil to help cover short-term foreign exchange needs. Chapter 7 shows that crude exports could reach significant proportions in the case of Japan and modest oil exports to the United States might eventually follow American recognition of Peking. Still, even if the "self-reliance" approach is now relaxed to a much greater extent than in the past, most indications as of 1977 suggested that the basic thrust of China's development strategy would remain unchanged. To the extent that oil exports grow, this is likely to reflect political as well as economic objectives, especially Peking's ambitions as a leader of developing countries in Asia and elsewhere. Peking may liberalize its foreign trade policies to meet specific, immediate objectives, but it continues to affirm its long-term goal of maximizing the self-contained strength of the Chinese economy and thereby limiting the future need for raw material exports.

Oil development has become attractive to Chinese planners precisely because it helps to make such an inward-focused strategy workable. The coal-to-oil shift is not only a logical response to the cost factors and changing domestic development priorities discussed earlier but also helps to create the type of economy that will best serve the goals of "self-reliance." By building a petrochemical industry China can convert some of its oil into exports of synthetic textiles and other finished products, in addition to meeting a variety of domestic needs. By using synthetics to reduce domestic cotton textile consumption, China can free some of its cotton textiles for export or, alternatively, release land hitherto devoted to cotton for grain production, thus reducing the necessity for grain imports. By using oil to manufacture fertilizer and to fuel agricultural mechanization, China can further relieve the need for food imports by stepping up food production. Increased domestic fertilizer production also means a reduced demand for fertilizer imports, and agricultural mechanization has the indirect benefit of releasing peasants for work in industry. The development of oil-based industries would clearly be more compatible with nationalist priorities than letting others use a major portion of Chinese crude to fuel their industries. Moreover, oil power strengthens China's military freedom of action and thus has a special significance in nationalist eyes. Domestic political and economic pressures reinforce each other, dictating a maximum effort to step up oil production as rapidly as possible.

The size of the investment required to keep up with these pressures would be so substantial that some observers have questioned whether China could muster the necessary resources even if oil development were given an overriding budgetary priority. Economist C. Y. Cheng, who predicted annual production of 335 million tons (2.5 billion barrels) by 1985 in the Commerce Department study cited earlier, also estimated that a $4.5-billion annual investment would be required to achieve this level.[29] Cheng took it for granted that this expenditure would be made. However, Randall W. Hardy, in a Federal Energy Agency study, cited even higher investment estimates and argued that outlays of such great magnitude for a single industry would be unlikely. Building on Chinese economic statistics published during the 1950s, Hardy constructed a model that he considered applicable to the present situation and estimated that a 20 percent annual growth rate in oil production would require almost doubling the 3 percent share of national investment believed to have been allocated to oil during the 1971–75 period.[30]

One of the implicit assumptions in the Hardy model was that a significant amount of future investment would have to be in the form of Chinese foreign exchange expenditure for the import of machinery and equipment. However,

Cheng demonstrated that expenditures on machinery and equipment, both domestic and imported, accounted for only half of the total investment in petroleum development during the 1953–74 period. Of this amount, expenditures on imports dropped from 65 percent of the total spent during the first five-year plan (1953–57) to 25 percent during the period as a whole. Cheng projected a further drop to 20 percent during the next decade, which would mean a relatively modest $450-million-per-year foreign exchange outlay out of his projected $4.5-billion investment projection.[31] Even this figure may be high, since it represents a projection of past trends and may not give adequate weight to recent evidence of greatly stepped-up efforts to establish indigenous equipment-manufacturing facilities. As the Chinese learn how to make more of their own equipment, the $2.25-billion-per-year figure suggested by Cheng for domestic machinery and equipment production could well increase in relation to expenditures on imports. In any case, $450 million per year would not necessarily be beyond Chinese capabilities, even within the framework of a relatively modest oil export program and a cautious approach to foreign credits. Similarly, in assessing whether China can afford the total of $4.05 billion in domestic outlays for oil development out of the overall $4.5 billion projected by Cheng, it should be remembered that attempts to quantify capital requirements in China on the basis of Western experience are inherently open to question on political and sociological grounds. Cheap manpower is available to Chinese economic managers on a vast scale, permitting exploratory drilling and the multiplication of new oil fields on a scale and at a pace that would not even be contemplated in a private venture. In more capital-intensive sectors of the petroleum industry, the political discipline and motivation of both skilled and unskilled labor are equally pertinent and cannot readily be measured by the budgetary yardsticks of Western industry.

As a totalitarian state, China can allocate its human and other resources without the same concern for an early

payoff that must necessarily govern the deployment of resources by Exxon or Gulf. Political and strategic considerations can be invoked to justify arbitrarily higher allocations
for petroleum than for other sectors, despite the necessarily
speculative character of oil exploration. At the same time,
there are finite limits to what any society can do. It remains to
be seen how the need for funds at any given period in such
lagging sectors of development as steel, transport, and agriculture will affect the rate of expansion in petroleum production. Given the interdependence already noted between
growth in petroleum and growth in these lagging sectors, a
slowdown in steel, in particular, could mean reduced output
by the petroleum equipment industry, just as cutbacks in oil
production could directly affect the rate of agricultural
mechanization.

The Technology
Question Mark

Among the American, Japanese, and European oilmen interviewed in the course of this study, the principal question at
issue in evaluating China's oil development was not whether
Peking could afford to make the requisite investments but
whether Chinese technical capabilities would prove adequate
in the absence of management contracts with Western oil
companies or other forms of comprehensive foreign assistance. The consensus was that the development process
would take longer without direct foreign help and would be
less efficient, but that China could make tolerable progress on
its own if it obtained enough indirect foreign assistance
through licensing arrangements and the employment of
foreign technical consultants. It was also strongly argued that
Peking would have to continue and even increase its already
substantial imports of specialized equipment and components from Western countries and Japan. Since 1972 China

has imported $110 million in petroleum equipment from the United States alone, including drilling bits, tubing, and blowout preventers made from high-grade alloy steels not yet produced in China. These components are essential for the manufacture of heavy rigs capable of drilling wells deeper than 10,000 feet, as well as for maximizing the efficiency of other drilling and production work. Should the factional argument between "radicals" and "moderates" lead to sweeping restrictions on such critical technological imports, it was felt, growth rates in oil production could drop sharply.

There can be no doubt that the degree of flexibility shown in the continuing implementation of the "self-reliance" policy will have a major bearing on how rapidly Chinese oil production increases. As it was evolving up to 1977, the "self-reliance" approach has been designed to avoid technological dependence, but not at the cost of shutting out needed foreign technology. What this means is that China seeks to defer the purchase of sophisticated foreign equipment or components from abroad until it has a nucleus of trained personnel capable of absorbing the technology involved for use within a Chinese-controlled administrative infrastructure. By 1972, when Peking began to buy Western and Japanese equipment, Chinese oil technicians had been working on their own for more than a decade following the departure of their Soviet advisers and had established the major elements of such an infrastructure. Subsequently, the terms of major purchases have reflected a growing Chinese self-confidence mixed with a cautious readiness for foreign help. In late 1975 Dresser Industries sold China $23 million worth of well-logging equipment (see chapter 4) involving highly complex electronic and other components. At Peking's insistence the Chinese technicians operating this equipment had only a six-week training program in Houston and one month of on-the-job training by Dresser advisers as part of the purchase contract, even though the Peking government knew it would have to pay disproportionately high rates for any technical help that might be needed at a later stage.

The foreign companies entering into such arrangements tend to believe that Peking will inevitably become dependent on them. In their view, China is so far behind the West in its oil technology that little risk is involved in selling sophisticated equipment to Peking. As for the Chinese, it is clear that they hope to use foreign equipment as prototypes for adaptation in their domestic manufacturing program. But it is also increasingly evident that some of the critical components needed to duplicate imported equipment cannot be produced in China at present. The most important of these bottlenecks are in offshore technology, as discussed in chapter 4, especially in the technology of deep-water drilling. Others affect both offshore and onshore operations, notably the need for high-grade alloy steels already mentioned. In addition to their use in heavy rigs, these high-grade alloys are critical in the manufacture of the large-diameter pipe needed for oil pipelines. China will also be peculiarly dependent on imports of computers and other highly specialized equipment for geophysical surveys and is not yet capable of making many of the miniaturized electronic components necessary to duplicate such equipment once prototypes are purchased. Significantly, foreign purchases of offshore geophysical survey boats were singled out in some of the more doctrinaire attacks on deputy premier Teng Hsiao-ping in early 1976,[32] but the defeat of the "Gang of Four" has once again strengthened the advocates of a flexible approach toward the import of petroleum equipment.

In evaluating just how much China may have to depend on imports of foreign technology, and for how long, it is useful to bear in mind that much of the progress being made in developing indigenous technical capabilities is shielded from view by the Chinese until they have something to show off. The West has repeatedly discovered how easy it is to underrate Chinese capabilities in the field of nuclear weapons development. In the case of oil, the West has similarly been largely unaware until recently of the concentrated efforts long under way in China to establish petroleum equipment factories and to overcome oil-related technical obstacles in metallurgy and other fields. When Soviet advisers left in 1960, China had only two large-scale petroleum equipment plants in Lanchow and a scattering of experimental factories elsewhere for the manufacture of drilling tools and pumps. By 1975 there were thirty-one major, identifiable petroleum equipment complexes and another seventy specialized factories known to the outside world.[33] When the Soviets left, China could make only a low-grade annealed steel; by 1964 Chinese factories were making fourteen types of alloys; by 1971 the number had gone up to eighty, and intensive efforts were continuing to narrow down the types needed from abroad, not only in oil-related enterprises but in all spheres of development. Japanese and other foreign visitors doing business with the Chinese oil industry have found indications of some twenty to thirty unannounced enterprises in the petroleum equipment field either already in operation or under construction.

By Western standards, "self-reliance" often results in crude products, but by all accounts these products may still be relatively serviceable. As one example, China still lacks the capacity to make some types of heavy-duty machine tools, which has made it necessary to fabricate petroleum equipment from a greater number of parts than would otherwise be necessary. This deficiency has slowed down the manufacturing process and has affected the durability and efficiency of the equipment, but the equipment has still been manufactured and used. Likewise, large computers would greatly speed up the processing of seismic survey data, but extensive surveys are being conducted anyway, through the painstaking use of less sophisticated equipment made in China's own rapidly developing computer manufacturing effort.

Announcing a stepped-up program of research and development on 8 May 1977, Petroleum Minister Kang Shih-en said that the "Gang of Four" had slowed down oil-related scientific progress by "purposely putting politics in opposition to professional work and being red in opposition to being expert." According to a New China News Agency account of his speech at a national "Learn from Taching" conference in Peking, Kang pledged a "big leap forward in the oil industry" based on a new policy departure:

The work of specialists must be combined with the mass movement for technical innovation and technical revolution. . . . We must develop scientists who are both red and expert. We must foster and select people for scientific and technical study from among workers with practical experience and run the petro-schools of various kinds well so as to keep sending a supply of new blood into the ranks of scientists and technicians. In this way, we can catch up and surpass the world advanced level in oil science and technology.

The Reserves
Question Mark

The most difficult question to evaluate in assessing China's oil potential is whether Peking can find new reserves rapidly enough to compensate for the depletion of existing fields that would result from a 12 to 13 percent annual rate of increase in production. Preponderant evidence suggests that China does have more than adequate onshore reserves to support a growth rate of this magnitude. At the same time, it cannot be taken for granted that Peking will be able to locate or develop *accessible* reserves in the face of significant geological and logistic obstacles. This chapter will conclude by summarizing what is known about onshore reserve prospects, in order to set the stage for a more extensive examination of the critical but elusive issue of offshore reserves.

Considering the present climate of optimism concerning Chinese reserves, both onshore and offshore, it is ironic to recall that China was widely regarded abroad as poor in petroleum until relatively recently. The Western majors thought of China as a market for their own oil, inspiring Hobart's 1933 best seller, *Oil for the Lamps of China*, a melodramatic saga of valiant Standard Oil salesmen who toiled in vain to light up the life of the Oriental heathen by popularizing kerosene lamps.[34] The majors made only desultory attempts to explore for oil in China prior to the Communist takeover and consistently reported unfavorable results. Even the lonely few in the West who argued that China might have significant unexplored deposits were talking of a mere 300 million tons (2.3 billion barrels) as late as 1960.[35] China now contends that Western—and Soviet—geologists deliberately downgraded their reserve estimates to discourage the development of Chinese oil resources and keep China dependent. Peking proudly points out that it was an innovative Chinese geologist, the late petroleum minister Li Su-kuang, who first recognized the vast oil potential of China and organized a greatly accelerated exploration program following the departure of the Soviet advisers in 1960. On the basis of Soviet geological data, Chinese recoverable onshore reserves were estimated to be only 2.7 billion tons (20.3 billion barrels).[36] By 1977 Peking had yet to announce official reserve estimates, but Chinese sources have informally referred to recoverable onshore reserves of at least 10 billion tons (75 billion barrels),[37] and offshore reserves could greatly increase this figure, as we shall see in chapter 3. Recoverable onshore and offshore reserves are estimated at 20.9 billion tons (157 billion barrels) for the United States and 68.4 billion tons (513 billion barrels) for the Middle East and the Persian Gulf combined.[38]

Estimates of recoverable reserves represent a percentage of the total reserves believed to exist on the basis of the available knowledge of potentially oil-bearing geological structures. However, the percentage used to determine recoverable reserves is not standardized and varies in accordance with differing geological assessments of global oil production experience.[39] Even in cases where experts have access to the same geological data, there is room for a wide divergence in recoverable reserve estimates, reflecting different estimates of total reserves and differing assumptions as to the recoverable percentage expected from those deposits. In the case of China, as in any closed society, the scope for controversy is still greater because the relevant geological data are only partially known to the outside world.

The first serious effort to estimate Chinese reserves was made by A. A. Meyerhoff, a respected American consulting geologist who has served as adviser to the Soviet Ministry of Geology with access to the Soviet geological data obtained in China from 1949 to 1960. It was Meyerhoff who revealed in 1970 that Soviet findings supported a 2.7-billion-ton (20.3-billion-barrel) estimate. Subsequently, on the basis of additional geological information from Japanese, American, and Chinese sources, Meyerhoff has allowed for the possibility of a much higher upper limit. Recoverable onshore reserves are "not less" than 2.7 billion tons (20.3 billion barrels), he wrote in 1975, but "may reach" 5.3 billion tons (40 billion barrels), exclusive of shale oil.[40] Meyerhoff's estimate is still regarded as much too low by a CIA analyst, Bobby A. Williams, who has concluded that recoverable onshore reserves should be estimated at not less than 5.9 billion tons (44.3 billion barrels) and "could easily" be 7.6 billion tons (57 billion barrels). Williams has supported these estimates with a detailed critique of Meyerhoff's implicit assumptions concerning the relationship between output and estimates of "proved" reserves.[41] In particular, Williams has stressed that Meyerhoff understated the probable rate of new reserve discoveries and relied excessively on outdated Soviet and Nationalist Chinese data.[42]

In general, foreign observers of the Chinese oil scene have been more cautious than Williams. A representative Japanese estimate set an onshore recoverable figure of 4.5 billion tons (33.8 billion barrels).[43] Three leading U.S. oil industry geologists have presented an onshore recoverable estimate (2.5 billion tons, or 18.7 billion barrels) slightly lower than Meyerhoff's minimum estimate of 2.7 billion tons (20.3 billion barrels).[44] Unpublished rough estimates by other majors tend to fall between 4 billion tons (30 billion barrels) [45] and

6.6 billion tons (50 billion barrels),[46] with the notable exception of one estimate (10.1 billion tons, or 75.8 billion barrels) [47] that is almost identical to the figure put forward by Chinese sources.

The possibility that most Western estimates are low must be seriously considered for two reasons. One is the Chinese tendency to shield knowledge of new reserves from the outside world until long after their discovery. The existence of the Takang, Shengli, and Chin Chou fields in the northeast and the Nanhai and Sansui fields in the southeast was not revealed for two to eight years after they went into operation. Four new fields at Fu-yu, I-tu, P'an-shan, and Chien-chang were producing 6 million tons (45 million barrels) in 1973, but this was not apparent to foreign observers until three years later. In 1976 it became clear that as much as 19 million tons (142.5 million barrels) of China's 1974 production had come from undisclosed fields in the east and northeast.[48]

A second and even more basic reason for taking Chinese claims seriously is that Peking has significant geological capabilities of its own and believes that conventional Western geological thinking impedes an adequate understanding of the distinctive tectonic factors present in China. Chinese scientists have long challenged classic geological concepts in which oil prospects are evaluated primarily in terms of marine sedimentary deposits, arguing that in China, oil is more likely to be found by studying the geophysical impact of prehistoric tectonic movements on inland basins and lake beds. In 1935 Li Su-kuang, then a professor at Peking University, first advanced this view in a series of lectures in England. His thesis was spelled out shortly thereafter in a book long forgotten by Western geologists but still regarded as a bible by the generation of Chinese geologists who were trained under his leadership. Pointing to the "clear development in China of certain outstanding tectonic types that are not always clearly revealed, for one reason or another, in other continental areas," Li, writing under the Anglicized name J. S. Lee, attacked the:

narrow channels of logic that do not always take sufficient heed of the arrangement of the ancient and existing mountain ranges brought about by orogenic forces. An investigation in which we are primarily concerned with geotectonic phenomena has sometimes curiously left tectonic geology out of account.[49]

His book was a forerunner of recent Western literature in the field of "plate tectonics," also known as sea-floor spreading or continental drift, which holds that the earth's crust is divided into a dozen or more large plates defined by earthquake belts.[50]

In conversations with Japanese visitors following the normalization of relations with Tokyo in 1972, Chinese geologists stressed that their oil deposits were not of marine origin.[51] The tectonic approach was also emphasized by Chinese geologists in a 1975 colloquium at the Lamont-Doherty Geological Observatory of Columbia University. Yen Tun-shih, deputy director of Peking's National Institute of Petroleum and Chemical Engineering, presented a paper at the colloquium that provided the most detailed exposition of Chinese geological thinking so far made available to a specialized audience outside of China since the Communist rise to power.

In this unpublished paper, obtained from a participant in the colloquium, Yen sought to show that Asia had its geological genesis in the ancient collisions of at least three smaller, previously existing continents. As a result, he explained, China is divided into two basically different regions. The western region is distinguished by high, northwest-trending ranges and basins presumably formed by northward pressure from India. The eastern region, more important geologically, bears the marks of the compression that resulted from the northwest thrust of the Pacific floor against China. This process occurred, he said, during the same geological epoch when the floor of some long-vanished ocean separating India from Asia was pressing in a northeasterly direction against China. The compression submerged what had once been saline lakes during the days of the dinosaurs, and beneath these lakes the deepest of China's oil reservoirs were formed. Conversion of organic material into oil was accelerated by volcanic heat generated during the compression period. In some cases, Yen said, the lakes were submerged to "unusually great" depths, producing oil-bearing formations more than three or four miles thick.[52]

Yen gave the name Pacific Movement to the "overall process of tectonic activity that has occurred in East Asia since the Meso-Cenozoic period." He pointedly observed that it was Li, "a brilliant Chinese geologist," who had formulated an original Chinese theory in which previous Western concepts were linked with the interaction of geotectonic "shear forms" and "east-west tectonic zones." [53] Ignored outside China, Li's concept "has been proved and enriched by oil prospecting practice," Yen said, and has recently been bolstered by a study of limestone samples collected by Chinese geologists from the summit of Mount Everest in May 1975. Chinese analyses showed that the limestone there had been laid down between 410 and 575 million years ago, substantiating the theory that northeasterly pressure against China from the floor of an ancient ocean had molded the Chinese tectonic environment.

While Yen did not make specific reserve estimates, he held up a glowing picture of an enormous oil potential and a successful oil development effort made possible by the application of Li's thinking. The Pacific Movement had defined the patterns of the "regularly arranged" sedimentary basins in eastern China and on the continental shelf, he said, and

In some areas bordered by many volcanic rocks, new petroliferous provinces have actually been found by tracing along the northeasterly trend of the basins. Even some of the minor fault basins should not be neglected, because they might contain very thick source beds and reservoir rocks.

In addition to determining the distribution of the basins, he added, "the tectonic stress of the 'Pacific Movement' also directly influenced the basement structures within the basins" and the manner in which oil and gas had migrated and accumulated. Where the compression was greatest, the subterranean landscape became a series of "islands" jutting up within deep basins. These "highs" provided natural repositories for limestone, creating "sedimentary basins with rich and varied source material" in which the thickness of deposits was "immense in magnitude":

This, coupled with the volcanic activities and the rise of the geothermal temperature of the area, improved the conditions for the conversion of organic materials into oil . . . and the accumulation and preservation of oil and gas in traps.

Precisely because of this geological history, he explained, the oil-bearing structures in eastern China and its offshore areas are extremely complex ("block-faulted quite intensively") in the very areas where they are deepest and most promising ("15,000–21,000 feet or more"). Nevertheless, on the basis of exploratory drilling to date, Chinese geologists have evolved a theory of "composite oil and gas accumulation zones" in which they view these structures as a series of "buried hills arranged in an echelon pattern" [54] and search for them accordingly.

The Chinese emphasis on tectonics is no longer disputed by Western geologists, and the Chinese discoveries on Mount Everest have been reinforced by the parallel findings of American scientists who have studied China topographically from space as part of international research efforts now under way in plate tectonics.[55] The earthquake analyses and images obtained by the Earth Resources Technology Satellite have provided new evidence of continuing tectonic pressure in central Asia, including indications that a cumulative shift of the earth's crust extending over 250 miles may have occurred along the Altyn Tagh Fault in mid-China.[56] As most Western petroleum geologists see it, however, the type of geological environment described by Yen is not as favorable as it might seem because the complexity and depth of the deposits make it inordinately expensive to produce oil.

Geological "traps" containing petroleum fall into two

categories—structural and stratigraphic. Structural traps, the type generally found in the Middle East, can be located readily from the surface by seismic instruments and often consist of big reservoirs. Stratigraphic traps, by contrast, have no surface reflection, are difficult to locate without drilling, and are less likely to yield big reservoirs. Although the Chinese themselves do not acknowledge it and offer some evidence to the contrary, most Western geologists believe that most of the oil found at Taching, Takang, and Shengli has been in stratigraphic traps. Maurice J. Terman, a U.S. Geological Survey China specialist, regards the northeast China fields as analogous to the Uinta Basin in northern Utah, an area that has been given a very low priority by the U.S. oil industry.[57] Finding oil in basins of this sort requires a well density considered to be economically questionable by Western oil companies, especially since the amount of oil ultimately recoverable from such fields can prove to be low by world standards. At the same time, while stratigraphic traps are more difficult to locate than structural traps, experience in the North Sea and Prudhoe Bay powerfully demonstrates that they can, on occasion, contain enormous quantities of oil. Necmettin Mungan, a leading Canadian petroleum engineer who has visited Taching, points out that analogies with the United States are misleading because stratigraphic traps tend to be much smaller in the United States than in China. Mungan believes that Terman "greatly underrates" the Chinese potential, possibly by three or four times.[58] Paul H. Fan, professor of geology at the University of Houston, an American of Chinese descent, also disputes the Terman analysis. Drawing on correspondence with former classmates now in key oil exploration posts in China, Fan reports that Chinese geological findings belie the Uinta comparison. Northeast China fields are more comparable to the Middle East, he states, given the extensive presence of limestone there.

China has made no attempt to conceal the formidable obstacles presented by the geological conditions at the fields in the northeast, especially Takang.[59] But the triumphant description of how these obstacles have been overcome [60] suggests that Peking may well be prepared to use its manpower and other resources in a fashion that Shell and the other companies involved in Uinta would not be willing to do. The reasons for this difference in attitude lie in the depth of the nationalist feeling in China discussed earlier and the role of oil in maximizing "self-reliance." Oil development is not designed to serve economic development as an end in itself; rather, it is addressed to economic priorities defined in relation to strategic and political goals. Even so, one should not underrate the scope of the challenge that China faces in seeking to find the reserves necessary to reach a 400-million-ton (3-billion-barrel) production level by 1990. Over time, the technical obstacles resulting from the nature of the Chinese geological environment will not necessarily prevent major discoveries, but these obstacles could significantly slow down the development of new fields and the expansion of existing fields in the years immediately ahead.

Why China
Must Go Offshore

Peking's problems are compounded by the regional distribution of its onshore reserves. The most convenient region in which to expand onshore oil development is eastern China, where existing production is concentrated and where there is ready access both to transportation and to already established industrial areas. As it happens, however, the largest portion of onshore reserves is believed to be in inaccessible interior regions close to the Soviet border. Most foreign observers estimate that as much as 65 to 75 percent of China's onshore reserves is located in such interior regions as the Dzungarian, Tarim, Szechwan, and Tsaidam basins (Figure 2). That this proportion may prove to be high was suggested by the emphasis given to eastern reserves by Chinese participants in the Lamont-Doherty colloquium and by other Chinese sources. Nevertheless, the interior regions are likely to have a greater potential than other onshore areas.

Faced with this regional distribution pattern, Peking will be compelled to make difficult choices concerning the allocation of investment capital, both between its eastern and interior areas and between onshore and offshore development. The urgency of these choices and the intensity of the pressures now building up to find new reserves has been illustrated by Randall Hardy in a hypothetical projection of what would happen if Peking should attempt to expand its oil production by relying solely on its reserves in eastern China. Hardy assumed a 2.7-billion-ton (20.3-billion-barrel) reserve base in the eastern region and the adoption of the standard reserve-to-production ratio of thirty used by most Western oil companies, i.e., Peking's adjustment of its production at any given time so as not to exceed one-thirtieth of the reserves then believed to exist. On this basis, new reserves would have to be found in eastern China at a rate of 400 million to 670 million tons (3 to 5 billion barrels) per year to sustain a 13 percent annual national growth rate in oil production between 1976 and 1983. In the absence of new discoveries, a growth rate of this magnitude would exhaust a 2.7-billion-ton (20.3-billion-barrel) reserve supply in eastern China soon after 1990.[61]

Viewed in terms of annual production levels, a 13 percent growth rate would mean an annual oil output of 200 million tons (1.5 billion barrels) by 1983. Yet, interviews with geologists and petroleum engineers suggest that the combined production of the Taching, Takang, and Shengli fields is not likely to reach 100 million tons (750 million barrels) before 1986–88 and is not likely to go much beyond this level. For one thing, the precise extent of the reserves in these fields is uncertain. For another, China has begun to shift to new water-injection methods that slow down production increases in order to maximize long-term recovery. The temporary drop in the production growth rate at Taching to 11.3 percent during the first quarter of 1975 was attributable to this shift.[62] In order to sustain a 13 percent growth rate, in short, it would clearly be necessary for Peking to complement the production of its northeast fields with an expansion of programs already under way to develop petroleum resources in other areas. This central reality points not only to the development of new fields in hitherto undeveloped areas of eastern China but also to stepped-up development of offshore areas or the interior regions or, more probably, a concerted effort on both fronts.

In his wide-ranging statistical study, Kim Woodard has powerfully demonstrated just how much difference off-shore development could make in China's petroleum future. If one rules out offshore development and assumes onshore reserves of only 5 billion tons (37.5 billion barrels), China would produce only 202 million tons (1.5 billion barrels) of crude in 1988 and its energy consumption would be geared to a 3 percent economic growth rate. But if one assumes, among other variables, total onshore and offshore reserves of 20 billion tons (150 billion barrels) and a major offshore effort, China would produce 527 million tons (4 billion barrels) in 1988 and would have a growth rate of 5 percent.[63]

The fundamental political and strategic considerations that make offshore development attractive are discussed at length elsewhere in this book. Even if these factors were not present, however, economic imperatives alone would in all likelihood have compelled Peking to undertake a significant offshore program. As chapter 3 shows, China's offshore reserves may well be as extensive as its onshore reserves. In the short-term future, there is also a strong economic case for a shallow-water offshore program as a less expensive means of increasing oil production than the development of interior fields, which would entail enormous indirect logistical and transportation costs. Dispersed as they are over vast distances, onshore reserves in the interior regions could be usefully exploited only in tandem with the construction of new railroad facilities or costly pipeline networks. These pipeline networks would have to be much longer than those needed for close-in offshore operations, since the offshore reserves in the Po Hai Gulf, the Yellow Sea, and the northern part of the

East China Sea are all located relatively close to the major centers of industrial energy consumption as well as the ports serving China's principal export customer, Japan. Moreover, rough terrain and the harsh climate would make it more difficult to install the type of pipeline required for the interior regions than those needed for close-in offshore activity. To be sure, offshore exploration and production operations are generally more expensive than onshore operations, especially in water over 250 feet deep. But China's initial offshore activity has centered in the Po Hai Gulf, where the water depth averages 66 feet, and in the Yellow Sea, where the average depth is 125 feet. For at least a decade China's major offshore production investments are likely to be concentrated in shallow-water areas, with deep-water activity steadily growing in importance but given a secondary priority.

In a long-term perspective, the cost factor is likely to have a diminishing influence on the relative emphasis given to offshore and interior activity. As its overall economic development progresses, China will acquire a nationwide transportation infrastructure that will open up expanding options for the oil industry along with other industries; and as offshore exploration moves into deeper waters, it will become progressively more expensive than the search for oil in the interior. At the same time, the pressure to find new reserves is likely to intensify as the years go by, giving enhanced importance to the vast offshore potential. Offshore development will become increasingly urgent, if only because China will need all the oil it can get to keep pace with its expanding development needs.

CHAPTER THREE

Another
Persian Gulf?

Is there "another Persian Gulf" somewhere in the East China Sea, the Yellow Sea, the Po Hai Gulf, or the South China Sea? Oceanographer K. O. Emery first suggested this analogy on his return from a series of United Nations seismic survey missions between 1968 and 1970, promptly touching off the vigorous competition that has since developed over offshore resources in East Asia.[1] As exploration has progressed, optimism regarding reserve prospects in the area has steadily intensified, with Emery's hopeful appraisal of the East China Sea equaled or surpassed by Japanese and other predictions focused, in particular, on the Po Hai Gulf and the waters near the Senkaku Islands (Tiao-yü T'ai).[2] Skeptics caution that exploratory drilling is still in its early stages, and geologists differ significantly regarding the possible size of the reserves; but there is a widespread and growing consensus that the East Asian offshore areas constitute one of the most promising of the unexplored undersea regions in the world.

The Numbers Game

The few attempts that have been made to quantify the off-shore reserves adjacent to China explain why offshore development looms so large to Peking as an answer to the economic dilemmas discussed in chapter 2. Depending on the outcome of continental shelf disputes with neighboring countries and the scope of the offshore areas ultimately exploited by Peking, the addition of an offshore dimension could at least double the Chinese oil potential. Even A. A. Meyerhoff, with his relatively low estimate of a maximum of 5.3 billion tons (40 billion barrels) for onshore reserves, has suggested an offshore figure of 4 billion tons (30 billion barrels).[3] This is broadly similar to offshore estimates by Soviet geologists,[4] by a Norwegian oceanographer, Jan-Olaf Willums,[5] and by most of the major oil companies.[6] However, it is considerably lower than some other Japanese and American estimates.

Meyerhoff divides his total into 1.7 billion tons (12.84 billion barrels) for the East China Sea, 1.1 billion (8.03 billion) for the South China Sea (inclusive of the Taiwan Strait), and 747 million (5.6 billion) each for the Yellow Sea and the Po Hai Gulf. By contrast, many Japanese estimates go as high as 10 billion tons (75 billion barrels) for the Po Hai Gulf alone. Ichizo Kimura, president of the International Oil Trading Company, told me that Chinese officials have estimated the Po Hai reserves at 8 billion tons (60 billion barrels), and Take-hiko Tominaga, managing director of Idemitsu Kosan, who led an expert mission of Japanese oilmen to China, found that "5 to 10 billion tons" (37.5 to 75 billion barrels) of recoverable reserves are believed to exist in the Po Hai Gulf. A Japanese government survey mission in 1969 suggested a 15-billion-ton (112.5-billion-barrel) figure for an undefined portion of the East China Sea shelf surrounding the Senkaku Islands (Tiao-yü T'ai).[7] Ted C. Findeiss, managing director of Clinton International, said that seismic studies showed a potential of 4 billion tons (30 billion barrels) in his Taiwan concession alone, 1.7 billion (12.8 billion) in one controversial structure that created the explosive political complications discussed in chapter 5.[8]

Meyerhoff excludes from his estimates potentially rich areas that are implicitly defined as part of its shelf jurisdiction by China, as we shall see, but have not been treated as such by the United States in United Nations Law of the Sea discussions. Other areas in the South China Sea, claimed by China for different reasons, have also been excluded. In any case, as the North Sea experience suggests, the full dimensions of the recoverable Chinese offshore potential could prove to be much higher (or lower) than 4 billion tons (30 billion barrels). Prior to drilling, the highest North Sea estimate was 2 billion tons (15 billion barrels). Later company estimates have ranged from 2.3 to 2.9 billion (17 to 22 billion).

Peking's own estimates of its offshore reserves have been stated only in broad, general terms. In late 1973 a scientific journal, noting that China had an 8,700-mile coastline and a continental shelf constituting one-twentieth of the world's continental shelf terrain, observed that "it is therefore not difficult to imagine that these immense continental shelves contain rich offshore oil reserves." The article added that "the offshore sedimentary basins lying on the continental shelves between the Tiao-yü Islands located at the farther end of the East China Sea and the coast of the China mainland are considered to be most favorable to oil prospecting and exploitation. Without any doubt this could lead to the discovery of a significant number of oil and gas fields in the future." [9] This article was followed by 1974 reports of "initial exploration" showing that "the continental shelf in China's seas has rich oil deposits, opening up new vistas for our petroleum industry." [10] A pro-Peking Hong Kong Chinese newspaper announced "very rich" offshore deposits, "especially in the continental shelf of the East China Sea and the area from the Po Hai Gulf south to Tsingtao," [11] and Radio Peking told of the discovery of "several huge fields in offshore areas" as well as on land.[12]

In the absence of extensive offshore drilling, either by Peking or by neighboring countries, most expert discussion of the East Asian offshore potential has necessarily been in general terms. What is known outside China about the reserve potential of waters adjacent to the mainland has been largely based on seismic surveys and other geophysical studies conducted under the auspices of a United Nations agency, the Committee for the Coordination of Joint Prospecting for Mineral Resources in Asian Offshore Areas (CCOP); oil companies, big and small, mostly American, Japanese, and British; geophysical companies operating independently, mostly American, German, and French; and government oil enterprises in Tokyo, Seoul, Taipei, Saigon, and Hanoi. None of these surveys has resulted in the public disclosure of quantified projections. However, the CCOP has published a flow of generalized findings that has kept expectations high.

The Emery Report
and Its Critics

The most widely cited findings published by the CCOP have been a series of reports by K. O. Emery, whose knowledge of the ocean floor in offshore Asian areas dates back to his sea-bottom studies for the U.S. Navy during World War II antisubmarine operations. Drawing on wartime records, Emery first hinted of petroleum possibilities in a preliminary 1961 study with a Japanese collaborator, Hiroshi Niino, who made use of Japanese wartime sea-bottom studies as well as postwar Japanese oceanographic surveys.[13] By 1967 Emery and Niino were talking explicitly of a massive oil potential in a more extensive study [14] that took into account further bottom samplings on the continental shelf and the published findings of Soviet oceanographers who had surveyed the offshore areas adjacent to northern China shortly before the Soviet rupture with Peking.[15] Terming the East China Sea "one of the potentially most favorable but little investigated regions among the offshore regions of the world," Emery and Niino wrote that "by comparison with other continental shelves, one can predict that the chances of success are likely to be good after a well-managed program of geophysical and geological exploration has been completed in the East China Sea." [16] They added that "the most favorable province for future submarine oil and gas fields is a wide belt along the outer part of the continental shelf" west of the Ryukyu Islands, an observation that contributed to the Japanese desire for the return of Okinawa (see chapter 7).[17] At the same time they were frank to say that the prediction of the most favorable parts of the continental shelf for future oil and gas fields would be "risky" if made on the basis of only the "few rocks that have so far been drilled from the sea floor." [18]

In an effort to develop more definitive data, Emery and CCOP officials organized a more ambitious survey in the fall of 1968 with the cooperation of the U.S. Naval Oceanographic Office and geologists dispatched by the Japanese, South Korean, and Nationalist Chinese governments. Based on more than 7,450 miles of continuous seismic profiles and accompanying geomagnetic profiles by a CCOP team aboard the U.S. Navy's *R. V. Hunt,* the 1968 survey report provoked widespread interest with its announcement of a

high probability . . . that the continental shelf between Taiwan and Japan may be one of the most prolific oil and gas reservoirs in the world. It is also one of the few large continental shelves of the world that has remained untested by the drill, owing to military and political factors.[19]

The study also reported a "second favorable area for oil and gas . . . beneath the Yellow Sea" where three broad, interconnected basins had been detected, one centered near Korea and two near the mainland of China. The report said that two of these basins contain sediments 4,600 to 4,900 feet thick and

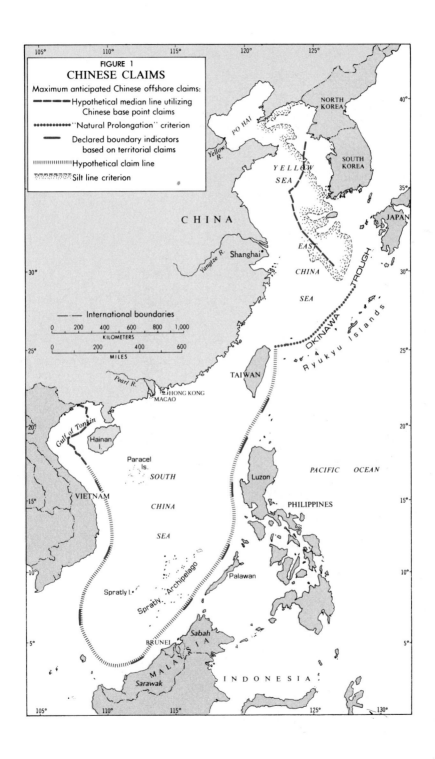

FIGURE 1
CHINESE CLAIMS

Maximum anticipated Chinese offshore claims:

━ ━ ━ Hypothetical median line utilizing Chinese base point claims

•••••••••• "Natural Prolongation" criterion

━━━━━ Declared boundary indicators based on territorial claims

||||||||||||||| Hypothetical claim line

Silt line criterion

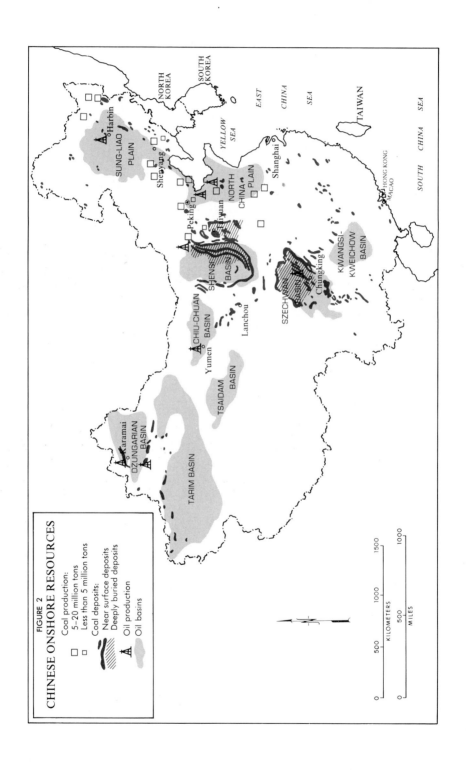

FIGURE 2
CHINESE ONSHORE RESOURCES

Coal production:
☐ 5-20 million tons
▫ Less than 5 million tons

Coal deposits:
▧ Near surface deposits
▣ Deeply buried deposits

▲ Oil production

Oil basins

TARIM BASIN

DZUNGARIAN BASIN
Karamai

TSAIDAM BASIN

CHIU-CHUAN BASIN
Yumen

Lanchou

SHENSI BASIN

SZECHUAN BASIN
Chungking

KWANGSI-KWEICHOW BASIN

SUNG-LIAO PLAIN
Harbin

Shenyang

NORTH CHINA PLAIN

Peking

Tsinan

Shanghai

NORTH KOREA

SOUTH KOREA

YELLOW SEA

EAST CHINA SEA

SOUTH CHINA SEA

TAIWAN

HONG KONG
MACAO

KILOMETERS
0   500   1000   1500

MILES
0   500   1000

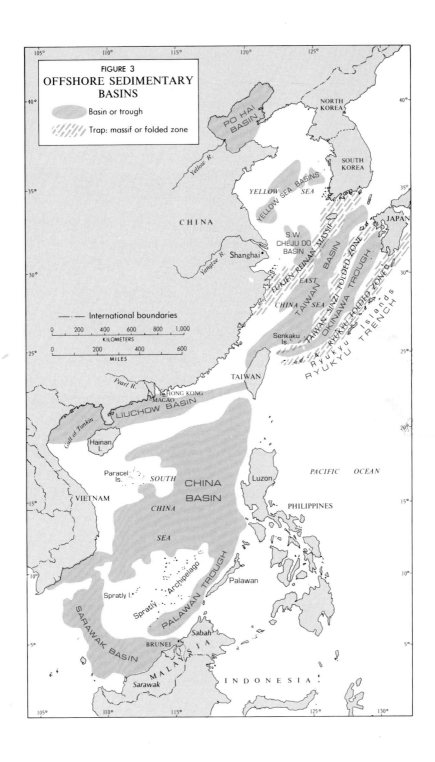

FIGURE 3
OFFSHORE SEDIMENTARY
BASINS

Basin or trough

Trap: massif or folded zone

105°   110°   115°   120°   125°

40°   40°

NORTH
KOREA

PO HAI
BASIN

Yellow R.   SOUTH
KOREA

35°   YELLOW   SEA   35°
YELLOW SEA BASINS

CHINA   JAPAN

S.W.
CHEJU DO
Shanghai   BASIN
Yangtze R.

30°   FUKIEN-REINAN MASSIF   30°
EAST
CHINA SEA   TAIWAN   BASIN
TAIWAN-SINZI-FOLDED-ZONE
OKINAWA TROUGH
RYUKYU-FOLDED-ZONE
Islands
Senkaku
Is.   Ryukyu
RYUKYU   TRENCH

International boundaries

0   200   400   600   800   1,000
KILOMETERS

0   200   400   600
MILES   25°   TAIWAN   25°

Pearl R.   HONG KONG
MACAO
LIUCHOW BASIN

Gulf of Tonkin
20°   20°
Hainan
I.

Paracel   PACIFIC   OCEAN
Is.   SOUTH   CHINA   Luzon
VIETNAM   BASIN
15°   CHINA   PHILIPPINES   15°

SEA

PALAWAN TROUGH
10°   Archipelago   Palawan   10°
Spratly I.
Spratly
SARAWAK BASIN
5°   Sabah   5°
BRUNEI   I A
MALA
Sarawak   INDONESIA

105°   110°   115°   125°   130°

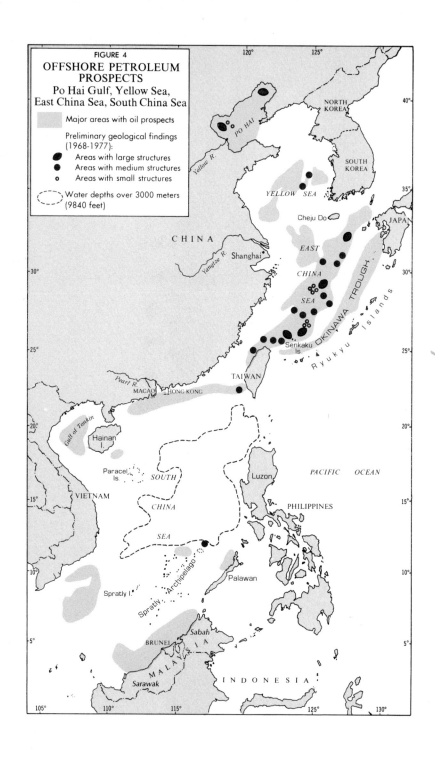

FIGURE 4

OFFSHORE PETROLEUM
PROSPECTS
Po Hai Gulf, Yellow Sea,
East China Sea, South China Sea

Major areas with oil prospects

Preliminary geological findings
(1968-1977):

● Areas with large structures
● Areas with medium structures
○ Areas with small structures

Water depths over 3000 meters
(9840 feet)

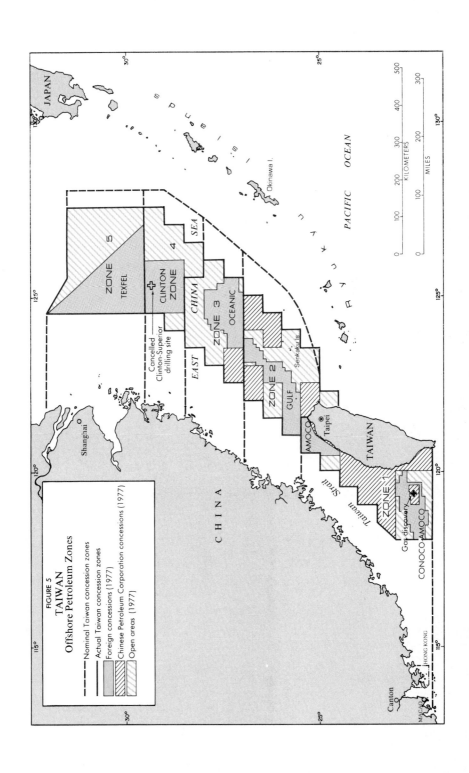

FIGURE 5
TAIWAN
Offshore Petroleum Zones

Nominal Taiwan concession zones
Actual Taiwan concession zones
Foreign concessions (1977)
Chinese Petroleum Corporation concessions (1977)
Open areas (1977)

JAPAN

Ryukyu Islands

Okinawa I.

PACIFIC OCEAN

KILOMETERS
0   100   200   300   400   500
0   100   200   300
MILES

125°

ZONE 5
TEXFEL

ZONE 4

CLINTON
ZONE

Cancelled
Clinton-Superior
drilling site

EAST   CHINA   SEA

ZONE 3
OCEANIC

ZONE 2
GULF

Senkakus

CHINA

Shanghai

120°

Taiwan Strait

AMOCO

Taipei

TAIWAN

ZONE 1

Gas discovery

CONOCO AMOCO

115°

CANTON

MACAU   HONG KONG

30°

25°

130°

25°

120°

30°

25°

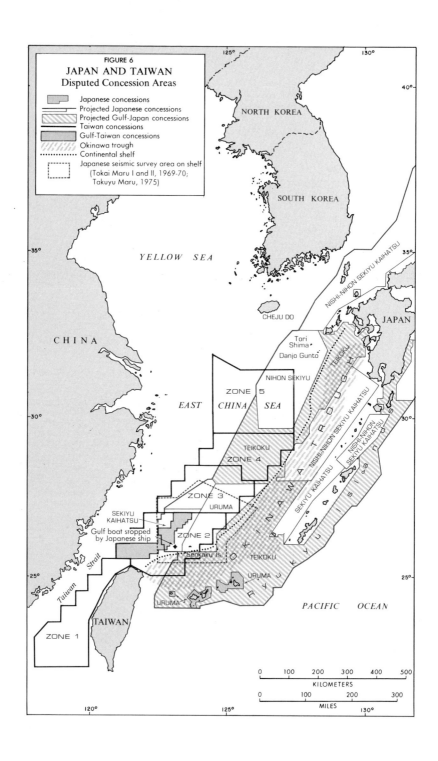

FIGURE 6

JAPAN AND TAIWAN
Disputed Concession Areas

Japanese concessions
Projected Japanese concessions
Projected Gulf-Japan concessions
Taiwan concessions
Gulf-Taiwan concessions
Okinawa trough
Continental shelf
Japanese seismic survey area on shelf
(Tokai Maru I and II, 1969-70;
Takuyu Maru, 1975)

NORTH KOREA

SOUTH KOREA

YELLOW SEA

CHEJU DO

CHINA

EAST CHINA SEA

Tori Shima
Danjo Gunto

NIHON SEKIYU

ZONE 5

ZONE 4

TEIKOKU

ZONE 3

URUMA

SEKIYU
KAIHATSU

Gulf boat stopped
by Japanese ship

ZONE 2

Senkaku Is.

TEIKOKU

URUMA

URUMA

TAIWAN

ZONE 1

Taiwan Strait

JAPAN

NISHI-NIHON SEKIYU KAIHATSU

TEIKOKU

OKINAWA TROUGH

NISHI-NIHON SEKIYU KAIHATSU

NISHI-NIHON
SEKIYU KAIHATSU

SEKIYU KAIHATSU

RYUKYU Islands

PACIFIC OCEAN

0   100   200   300   400   500
KILOMETERS
0   100   200   300
MILES

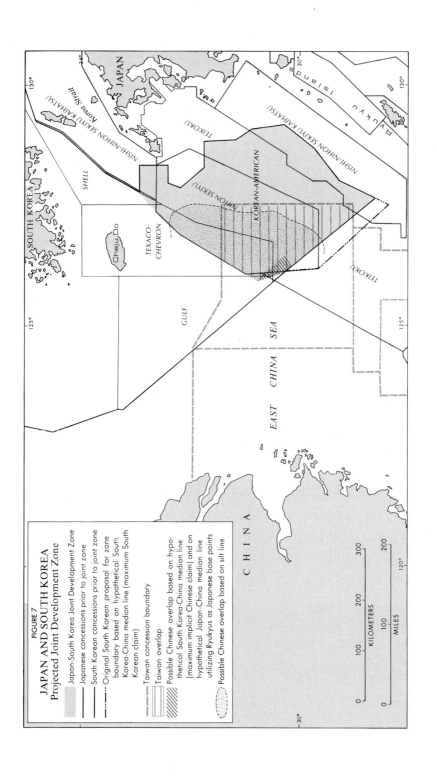

FIGURE 7

JAPAN AND SOUTH KOREA
Projected Joint Development Zone

- Japan-South Korea Joint Development Zone
- Japanese concessions prior to joint zone
- South Korean concessions prior to joint zone
- Original South Korean proposal for zone boundary based on hypothetical South Korea-China median line (maximum South Korean claim)
- Taiwan concession boundary
- Taiwan overlap
- Possible Chinese overlap based on hypothetical South Korea-China median line (maximum implicit Chinese claim) and on hypothetical Japan-China median line utilizing Ryukyus as Japanese base points
- Possible Chinese overlap based on silt line

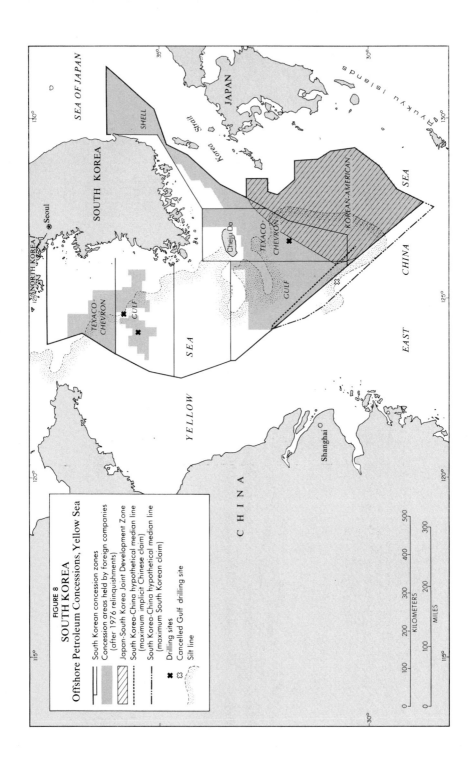

FIGURE 8
SOUTH KOREA
Offshore Petroleum Concessions, Yellow Sea

South Korean concession zones

Concession areas held by foreign companies
(after 1976 relinquishments)

Japan-South Korea Joint Development Zone

South Korea-China hypothetical median line
(maximum implicit Chinese claim)

South Korea-China hypothetical median line
(maximum South Korean claim)

Drilling sites

Cancelled Gulf drilling site

Silt line

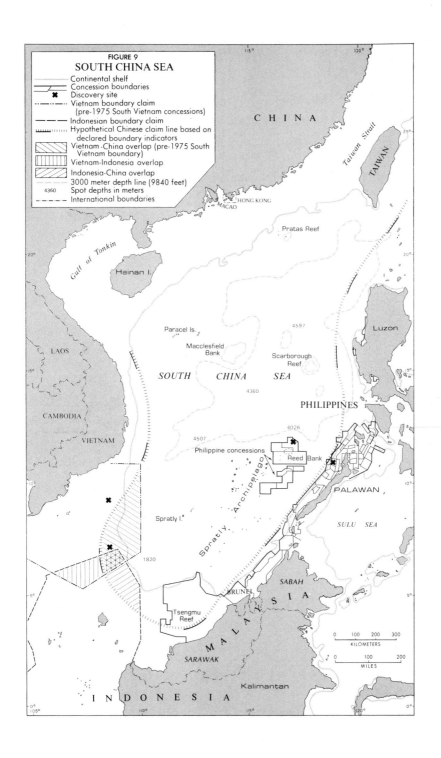

FIGURE 9
SOUTH CHINA SEA

———— Continental shelf
———— Concession boundaries
✖ Discovery site
—··—··— Vietnam boundary claim
(pre-1975 South Vietnam concessions)
———— Indonesian boundary claim
·····ıııı····· Hypothetical Chinese claim line based on
declared boundary indicators
▨ Vietnam-China overlap (pre-1975 South
Vietnam boundary)
▥ Vietnam-Indonesia overlap
▨ Indonesia-China overlap
———— 3000 meter depth line (9840 feet)
4360 Spot depths in meters
—·—·— International boundaries

C H I N A

TAIWAN

Taiwan Strait

HONG KONG
MACAO

Pratas Reef

Gulf of Tonkin

Hainan I.

Luzon

Paracel Is.

4597

Macclesfield
Bank

Scarborough
Reef

SOUTH   CHINA   SEA

4360

LAOS

CAMBODIA

VIETNAM

PHILIPPINES

4026

4507

Philippine concessions

Reed Bank

Spratly I.

Sulu Sea

PALAWAN

1820

Spratly   Archipelago

SABAH

BRUNEI

MALAYSIA

Tsengmu
Reef

SARAWAK

Kalimantan

I N D O N E S I A

0   100   200   300
KILOMETERS
0   100   200
MILES

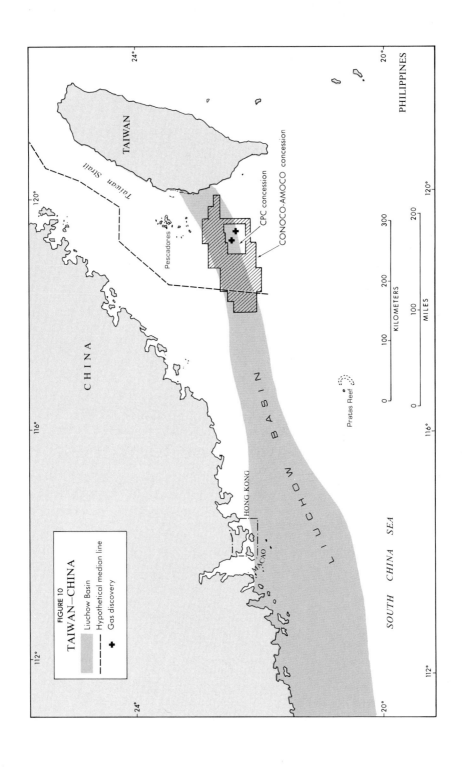

FIGURE 10
TAIWAN—CHINA

Liuchow Basin

Hypothetical median line

+ Gas discovery

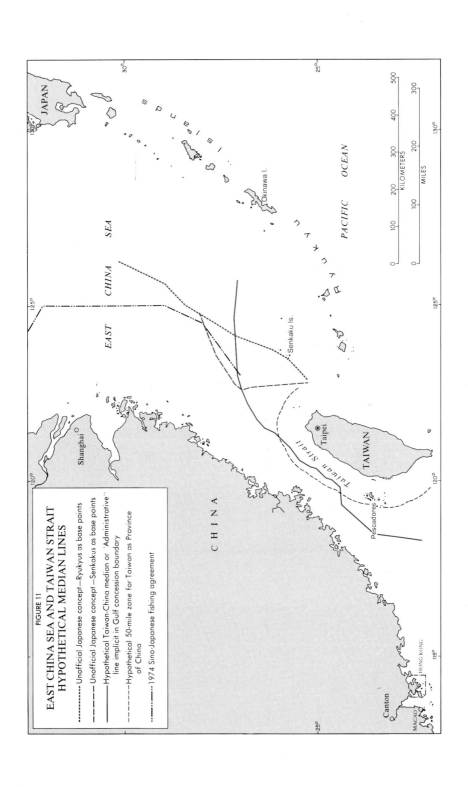

FIGURE 11

EAST CHINA SEA AND TAIWAN STRAIT
HYPOTHETICAL MEDIAN LINES

········· Unofficial Japanese concept—Ryukyus as base points
———— Unofficial Japanese concept—Senkaku as base points
———— Hypothetical Taiwan-China median or "Administrative"
         line implicit in Gulf concession boundary
– – – – Hypothetical 50-mile zone for Taiwan as Province
         of China
·–·–·– 1974 Sino-Japanese fishing agreement

JAPAN

EAST   CHINA   SEA

Shanghai

CHINA

Senkaku Is.

Okinawa I.

R Y U K Y U   I s l a n d s

PACIFIC   OCEAN

Taipei
TAIWAN
Taiwan Strait

Pescadores

Canton
HONG KONG
MACAO

KILOMETERS
0   100   200   300   400   500
MILES
0   100   200   300

Figure 12: A Chinese View: Offshore Oil And Taiwan

Comparative Estimates of Oil Deposits of China, the United States and the U.S.S.R. Before and After Discovery of Taiwan Basin

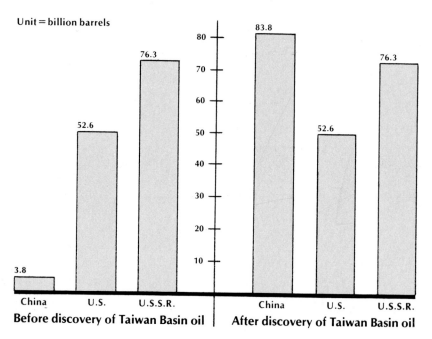

Unit = billion barrels

Before discovery of Taiwan Basin oil: China 3.8, U.S. 52.6, U.S.S.R. 76.3

After discovery of Taiwan Basin oil: China 83.8, U.S. 52.6, U.S.S.R. 76.3

Adapted from <u>Ming Pao</u>, an independent Hong Kong monthly, May, 1971.

a concentration of organic matter as great or greater than that likely to exist in the continental shelf sediments between Japan and Taiwan.[20]

The Emery report backed up its conclusions with a geophysical analysis underlining the importance of three parallel ridges that create natural receptacles for the deposit of sediment from the Yellow and Yangtze rivers in these basins (Figure 3). The first of these ridges, a subsea extension of the Lao Yehling mountain range known as the Shantung-Lao Yehling Massif, juts out from the Shantung Peninsula and traps the sediments from the Yellow River in the Po Hai Gulf. The second, known as the Fukien-Reinan Massif, crosses the mouth of the Yellow Sea, trapping at least 48,000 cubic miles of sediment there. Finally, the Taiwan-Sindzi Folded Zone, running along the eastern edge of the continental shelf, traps another 168,000 cubic miles.

As the Emery report points out, the Yellow and Yangtze rivers drain sediments from a vast area constituting nearly one-third of the total land mass of China. The Yellow River in particular is one of the most heavily silt-laden of the principal rivers of the world. Rising high in the western province of Tsinghai, it winds its way eastward for nearly 3,000 miles. Often breaching its dikes during floods, the Yellow River has radically shifted its course through the north China plain; entering the Yellow Sea south of the Shantung Peninsula prior to the mid-nineteenth century, floods later shifted its course northward. In 1938 the Chinese cut the dikes to divert the river southward again as a means of delaying the invading Japanese armies. In 1947 the dikes were rebuilt to return the river to its northern course. Such is the force of the Yellow River that its silt is extending the river's seaward march at an estimated rate of a mile per decade.

While the sediment deposited by the Yangtze is considerably less than that of the Yellow River, the Yangtze winds its way for an even greater distance of some 3,400 miles from the Tibet highlands, and its delta pushes outward at a rate believed to exceed a mile per century. The volume of sediment annually discharged from these two Chinese rivers has been estimated at 2.08 billion tons for the Yellow River and 550 million tons for the Yangtze. Emery stresses the "huge drainage areas and sediment loads" of the two waterways and estimates that the rate at which their sediment has been deposited is five times greater than that for the Atlantic continental shelf of the United States.[21]

Emery's estimates of the extent of sediment deposits—and thus the potential for oil-bearing structures—have been criticized as too conservative in some respects and as too optimistic in others. Maurice Terman, a U.S. Geological Survey China specialist, has argued that Emery is much too cautious in his estimates for the two subbasins in the western portion of the Yellow Sea Basin. While Emery has suggested the existence of sediments between 4,600 and 4,900 feet thick, Terman has estimated that the deposits in the Po Hai Gulf and the Yellow Sea may be "three or four times deeper" than that. At the same time, Terman attributes this thickness to the tectonic factors discussed in chapter 2, which leads him to an ambivalent judgment as to the feasibility of oil development in these areas. In Terman's view the geological characteristics of the Po Hai Gulf and Yellow Sea offshore areas have been determined by the same tectonic phenomena that have conditioned the onshore areas of north and northeast China.[22] In both cases he asserts that these are characteristic of a continental rather than a marine geological environment. Terman emphasizes that while this could mean large deposits of oil, the complexity of the resulting geological environment could make the location and extraction of the oil disproportionately expensive in offshore as well as onshore areas.

Terman's assessment of the Po Hai Gulf and the Yellow Sea as a continental rather than a marine geological environment is sharply contested by geologist Paul H. Fan, who maintains that Terman's analysis, based solely on the type of seismic data relevant for earthquake research, fails to take into account unpublished Chinese geological findings. According to Fan, Chinese studies have shown the Po Hai and

Yellow Sea areas to be marine geological environments. Japanese geologists have also been told of these studies and have been given samples of sedimentary rock said to be from the Takang field, adjacent to the Yellow Sea, which showed the field to be in a marine geological environment. However, on the basis of their exploration in the Yellow Sea, Gulf Oil Company geologists presented preliminary findings in 1975 that lend a measure of support to the Terman assessment while also stressing the magnitude of the oil reserves there. Pointing to three basins in which further exploration is warranted (Figure 3), the Gulf report notes that the so-called Western Subbasin contains more than 3.1 miles of sediments but is geologically complex and "deeply depressed by faulting." As for the Socotra Subbasin (the site of a potential dispute with China discussed in chapter 6), the report states that it includes a northwest-southeast strait in which the sedimentary deposits were found to be at least 2.2 miles in depth.[23]

Apart from the unresolved expert argument over the tectonic environment, many geologists have contended that Emery, as an oceanographer, also failed to take adequately into account other geological factors. By focusing solely on the volume of the sediments and failing to consider the nature of the rocks involved, said John F. Mason, former director of exploration for the Continental Oil Company (Conoco), Emery may have automatically overestimated the oil potential of the areas surveyed. In any case, Mason explained, the CCOP was inherently foreclosed from accurately determining the character of the rocks because its survey boat was not equipped to use as much dynamite as private oil companies normally use, and its seismic testing was thus unable to reach the required depths. C. Y. Meng, chief geologist of the Chinese Petroleum Corporation in Taiwan, stressed that the Emery survey showed sediments only up to 1.2 miles in thickness and that the sediments in many areas were actually "much deeper," which could mean an underestimation of the oil potential in those areas.

The tendency of some oil company officials to discount the Emery report reflects in part their coolness toward the very idea of a regional intergovernmental agency conducting public oil survey operations. Gulf, Amoco, and Western geophysical companies had already quietly begun modest preliminary surveys of their own in 1967, and the excitement generated by the Emery findings in the countries concerned was a source of great irritation at the time to many of the companies, which felt that the report complicated their bargaining in then pending negotiations for offshore concessions. Partly as a result of company pressures and partly as a result of bureaucratic rivalries in the U.S. Geological Survey, Emery was eased out as the principal U.S. representative in CCOP meetings. In their estimate of the offshore potential, however, most oil companies appear to share the positive appraisal reflected in the Emery report.

While many oilmen consider Emery's Persian Gulf analogy to be unduly optimistic, most agree with his overall judgment that the East Asian offshore areas are promising. Their technical data are closely held, to be sure, and their public posture does not generally correspond to their actual plans and intentions. Nevertheless, I was able to learn something of what oil people really think by interviewing exploration executives and geologists in Asia-minded oil companies in the United States, Europe, and throughout Asia as part of this study during 1974–76. I made a point of talking separately with officials working in different places within the same company and of checking their statements not only against one another but also against what they say to the governments concerned. In some cases, officials of leading American and Japanese companies made partial data available on a confidential basis after satisfying themselves that my research was not a cover for someone's commercial or other intelligence operations. In others, where companies holding concessions had attempted to sell an interest to other companies, I was able to see proprietary seismic data used during the negotiations. As my investigation proceeded, more and more doors opened, and it became clear that the offshore areas adjacent to China are appraised with a remarkable degree of unanimity in informed quarters. While there are differences with respect to specific areas, there is widespread agreement that these areas collectively constitute one of the most promising of the unexplored offshore regions.

The limited drilling conducted so far has been inconclusive and has had little effect on this appraisal. It is generally recognized that the extent of drilling to date has been negligible, by international standards, and that the most promising areas in geological terms have not yet been drilled for political reasons. In the case of the Yellow Sea, the most promising geological structures lie on the Chinese side of any future median line, and drilling in potentially contested areas in the middle portion has hitherto been too great a political risk (see chapter 6). Similarly, the most promising geological opportunity in any Taiwan concession appears to be located in the Clinton zone, where political considerations have also led to a cancellation of proposed drilling plans. In the one case of a significant discovery in the area, Conoco found gas off the southwest coast of Taiwan after relatively limited drilling but has been hesitant to pursue further exploration in view of the uncertain status of Taiwan.

In a paper submitted to the 1975 World Petroleum Congress, Mason and two Amoco geologists alluded to the Taiwan gas discovery as well as to successful exploration in the Po Hai Gulf and Southeast Asian offshore areas, concluding:

Relatively unexplored areas along the mainland coast from Indochina to Korea have reasonably good prospects . . . in spite of dry holes in the East China Sea north of Taiwan and the Yellow Sea off Korea. The results of exploration to date suggest that continental borderlands and shelves offer the best prospects for the discovery of additional large reserves of oil and gas within the region currently open to foreign operators, and that among these areas . . . the tract fronting the Pacific Coast of Asia is perhaps the most attractive.[24]

While Middle East oil is found primarily in limestone, with its rapid-yielding qualities, there is some evidence that the East China and Yellow seas may have sandstone or other nonlimestone reservoirs. If so, geologically speaking they would not represent "another Persian Gulf." They may prove to be more comparable to the North Sea, which is primarily sandstone but differs from the East China and Yellow seas in that it also has extensive shale source beds. Mason likens the potential of East Asian offshore areas to Wyoming; Rober E. King, former exploration director of Caltex, to Nigeria or Indonesia; A. A. Meyerhoff, to the eastern North American continental shelf; and Paul E. Ravesies, vice-president for exploration of Atlantic-Richfield, to the U.S. Gulf coast.

Unanswered Questions

The principal unresolved issue among geologists with respect to the East China and Yellow seas is whether they contain big reservoirs that would be easy to tap or complex, "faulted" structures. Seismic studies conducted by the Japanese government and some of the companies with Taiwan concessions suggest that certain parts of the continental shelf to the north and east of Taiwan are likely to have large reservoirs. However, Mason and his two Amoco colleagues said that while their studies show "reasonably thick" sediments in the area, "individual fields are not commonly in the giant category." [25] Geologists for another leading company said in an interview that with notable exceptions, such as the northeast corner of the Clinton zone (Figure 5), much of the petroleum abundance of the East China Sea appears to lie in scattered small accumulations. The amount of oil in each case might total between 3.3 and 13.3 million tons (25 and 100 million barrels), they added, which would be enough to warrant exploratory onshore drilling by a private oil company but might not be enough to justify the disproportionately high costs of offshore exploration.

Pointing to the "attractive petroleum potential" of the East Asian offshore areas, Maurice Mainguy, a French geophysicist formerly associated with the CCOP, observed:

The East China Sea Basin alone, extending from Taiwan to the north of Tsushima, is as big as the Persian Gulf Basin, and hopes for new producing areas under the sea are great. But the greater part of the most promising areas is under rather deep seas and far from the coast. The cost of producing oil under such conditions is high and, to be competitive with the Middle East oil, the fields to be discovered must be very big in size, and the productivity of individual wells, which is the key factor in determining the economic value of an oil field, must be high; this factor itself is linked to the quality of the reservoir rocks.[26]

Viewed together with the sea boundary disputes involved, the high costs of offshore exploration have loomed unusually large in the eyes of private Western operators approaching East Asia, where climatic conditions tend to be unfavorable. Both the Yellow and East China seas are noted for winter gales and summer typhoons that make for a relatively short drilling season from April to July. There is also a suspicion that gas is more likely to be found in the area than oil; this would be an added deterrent to private exploration, since gas can be much more expensive to exploit than oil. It would either have to be stored in liquefaction plants built on mid-ocean platforms or, alternatively, piped to shore, which could be inordinately expensive in cases involving long distances. By contrast, offshore oil can be piped to shore or loaded into tankers at midocean terminals, depending on the distance from shore and the topography of the sea bottom. The belief that the East China Sea may contain more gas than oil was expressed by Akinobu Tsumura, director of research for the Japan Petroleum Development Corporation, and by another Japanese government petroleum expert, Yasufumi Ishiwada.[27] Robert E. King contended in an interview that gas is more likely than oil because the sediments drained from Chinese rivers are predominantly composed of humic matter from plants, as against marine organisms, which has generally resulted in gas in most other areas consisting mainly of deltaic sediments. Mainguy, too, has written that "there is some evidence of the possible presence of more gas than oil along the continental shelf of eastern Asia" and stresses the "great value" of gas for industrializing countries in addition to the high cost of producing it offshore.[28] It should be noted here, however, that cost factors would not necessarily be viewed in an identical manner by foreign private enterprises and Chinese officials. As emphasized earlier, strategic and political considerations could lead Peking to make both exploration and production investments that would be regarded as uneconomic by a private company.

In the case of the South China Sea, exploration activity has been constrained not only by the cost and by the political uncertainties found in other East Asian offshore areas but also by the unusually deep water there. A CCOP task force found that the water depth in one part of the South China Sea, known as the China Basin, exceeds 2.5 miles in some places, which is far beyond the reach of present deep-water production technology. As the task force report noted, however, "more than half" of the China Basin is not that deep, and much of it is now or will soon be within the reach of drilling technology, including areas near the Paracel and Spratly islands believed to contain petroleum deposits.[29] In the long run, many geologists think that the deep-water reserve potential of the South China Sea holds even more promise than that of the East China and Yellow seas. In the immediate future, intense interest is likely to be focused on the area between Taiwan and Hainan, often identified as the Luichow Basin, as well as areas to the east and south of Hainan and the Tonkin Gulf.

Much of what is publicly known about the oil potential of the South China Sea is the result of a successor voyage by the *R. V. Hunt* following the 1968 survey that resulted in the Emery report. Cosponsored by the CCOP and the U.S. Naval Oceanographic Office, as was the first survey, the South China Sea reconnaissance cruise was conducted from 3 June to 27 August 1969. Like the initial Emery mission, it was followed by several private seismic surveys commissioned by oil companies and governments in the area.

In one analysis based on the 1969 survey data, Emery and his collaborator, Zvi ben-Avraham, concluded that the greatest oil potential in the South China Sea was to be found along the strip of continental shelf between Taiwan and Hainan, relatively close to the Chinese coast (Figure 3). At the same time, they added that a series of ridges cutting across the China Basin in the central portion of the South China Sea is of "even more speculative interest" and warrants further study. One particular ridge at the edge of the continental shelf is believed to have trapped sediment and is cited in support of optimistic predictions that a "filled marginal trough" underlies the continental shelf between Taiwan and Hainan.[30]

Another analysis based on the 1969 R. V. Hunt survey noted that there were "considerable expectations" for off-shore production in the Mekong and Brunei-Saigon basins to the south and southwest of Vietnam and that these expectations appeared to be "well based." [31] However, an earlier CCOP study of offshore prospects for South Vietnam stressed that the "most hopeful" area for discovery of offshore oil near Vietnam lay in the very part of the Brunei-Saigon Basin where the water depth posed significant technical problems. This analysis also mentioned the possibility of oil in the Paracels.[32]

In his ambitious attempt to quantify the petroleum reserves in specific East Asian offshore areas, oceanographer Jan-Olaf Willums has singled out the "deep-water sections around Hainan" as the most promising parts of the South China Sea, estimating an average hydrocarbon potential there of 51,800 to 59,600 barrels per square mile. Willums rates the overall potential of the South China Sea as "substantially lower" than that of the East China Sea and less promising than other offshore areas adjacent to China. In the deeper sections of the East China Sea shelf and the Taiwan Strait, he writes, the hydrocarbon potential surpasses 129,500 barrels per square mile, reaching 204,600 barrels in the area to the north and west of the Senkaku Islands (Tiao-yü T'ai); and in parts of the Yellow Sea and the Po Hai Gulf, the potential approaches 77,700 barrels.[33]

My own conclusion after the extensive interviews

mentioned earlier is that it would be premature to attempt volumetric estimates on the basis of the limited geological data so far made available by China and the limited exploration that has occurred in the concessions granted by neighboring countries. However, by piecing together published studies and numerous fragments of unpublished proprietary data, it is possible to suggest in rough outline where the most promising geological structures are believed to exist. Figure 4 does not purport to delineate these structures in the precise terms that would be useful to the geologist. Rather, it represents an impressionistic composite picture of the conventional wisdom among experts in the United States, Europe, and many of the East Asian countries involved regarding the results of seismic studies to date. As exploration progresses, this map will require continual amendment; but for the present, if viewed in relation to implicit and explicit national concession claims (Figures 5 to 9), it helps to define the most probable focal points of the incipient offshore resource conflicts discussed in this book.

Chapter Four

China Goes Offshore

It is one thing to conclude that China will grow increasingly hungry for offshore resources as its energy needs multiply and that the East Asian continental shelf does, in fact, hold large undiscovered oil and gas reserves. It is another and more difficult matter to evaluate how far and how fast China is likely to go in developing a viable offshore program within the constraints imposed by its "self-reliance" commitment. This chapter seeks to project the pace of Chinese progress in both exploration and production, not only because offshore petroleum development could critically affect Peking's economic potential but, more important for this analysis, because the state of Chinese offshore capabilities at any given time could directly govern how Peking handles its continental shelf disputes in political terms.

The Preparatory Phase:
1957—1974

In assessing Chinese offshore progress to date it is necessary to bear in mind that Peking has been sensitive to the potential of the continental shelf since the late 1950s. The Emery report and the appearance of Western survey boats accelerated a preparatory process already under way well before other East Asian coastal states and their foreign collaborators began to explore for offshore oil in the area. In 1958 and 1959 a leading Chinese oceanographer, Ch'in Yun-shan, published reports of the first marine geological surveys of the Po Hai Gulf, the Yellow Sea, and the East China Sea,[1] and in 1960 a Chinese counterpart to *Popular Science* announced the discovery of "convincing evidence" that justified serious prospecting for offshore oil. "The seas surrounding China cover one of the world's rare continental shelf areas," said the 1960 article in the Shanghai journal *Ko-Hsüeh Hua-pao*. Assessing the new marine geological findings in relation to earlier onshore geological analysis, the article said that already established onshore petroleum basins in northeast China extended into the Po Hai Gulf, the Yellow Sea, and the East China Sea, adding that "Po Hai, China's inland sea, is only several tens of meters deep and is thus especially suited to prospecting and mining."

The excitement generated by these newly aroused hopes for offshore riches was apparent in a Rube Goldberg-like discussion of possible ways that the resources of the sea bottom might be exploited in future years. The author brushed aside existing offshore rig technology with the assertion that it would be necessary to operate beneath the ocean surface in order to avoid the impact of storms. In a scheme suggestive of the subsea production systems that have since been developed in the West, he proposed a submarine fixed at a depth of 330 feet "as the main working station of the 'ocean-bottom miner'" to be utilized for both oil and gas production as well as the potential mining of deep-sea mineral resources such as manganese nodules. The article further explained:

When the sea is calm a hollow pipe can be raised to the surface of the ocean for air or for transporting personnel to and from the surface by means of elevators. The submarine working station can also obtain air from the water . . . and the submarine must be well protected against wind and waves. Automatic drilling equipment may be lowered from the submarine to the sea bottom and connected to the surface or to coastal oil or gas pipelines. A large remote-controlled submarine tank equipped with automatic drilling equipment could also be built. This submarine tank could drive from the coast to the ocean floor, and could pull behind it a tube to transport the gas or oil.[2]

Continuing their researches, Chinese oceanographers and geologists conducted a study of the coastal areas off Chekiang Province south of Shanghai in 1961 and a further study of the Po Hai Gulf in 1962.[3] In a lengthy 1963 study summarizing the principal findings of the surveys conducted up to that point, Ch'in Yun-shan stressed that the East China Sea continental shelf "is one of the widest continental shelf areas in the world." However, while clearly fascinated by the potential of the East China Sea, he focused attention on the more immediate opportunities for oil exploitation in shallower waters, emphasizing the contrast between the water depths of the Po Hai-Yellow Sea area, rarely exceeding 200 feet, and water three times deeper on the easternmost extremities of the East China Sea shelf. He presented a geological analysis of the interior areas of the continental shelf relatively near to China, with their finer sedimentation and greater deposits of organic matter, and exterior areas of the continental shelf with coarser sedimentation and less organic matter. The study contained extensive statistics concerning the distribution and character of the sedimentary deposits in each area as well as basic oceanographic data relating to width, depth, and ocean currents. Ch'in's bibliography showed a close awareness of foreign studies of the Chinese offshore areas, and he specifically challenged the analysis presented in the 1961 study by Hiroshi Niino and K. O. Emery,[4] stressing instead the different stages at which the interior and exterior sediments were deposited. Ch'in pointedly emphasized the "intensive accumulation" of sedimentary deposits in the Po Hai and Tonkin gulfs made possible by the level sea floor in both of these offshore areas.[5]

Even before Soviet technical advisers left China in 1960 as a result of the Sino-Soviet split, Peking had begun to make selective imports from non-Communist countries related to its oil development program. Thus, Enrico Mattei, creator of the Italian state oil enterprise, Ente Nazionale Idrocarburi (ENI), visited China in 1958 and concluded a controversial $200-million trade deal in 1961 prior to his death in a 1962 plane crash.[6] While the bulk of this deal focused on refineries and refinery equipment, there is reason to believe that other equipment related to onshore and offshore oil exploration was included in ENI transactions even at this early stage. It should also be remembered that China's ties with Romania date back to the early 1960s and included collaboration in oil development from the start. Similarly, the gradual change in French policy toward China in the early 1960s resulted in some sales of oil-related equipment, including a marine navigation system suitable for use in offshore oil surveys.

Little is known concerning the full extent of Chinese exploration activities during the 1960s, especially during the Cultural Revolution, but most evidence suggests that the Chinese fleet of some thirty-five oceanographic vessels has included at least three marine geological ships for more than a decade. Given Chinese progress in seismological and computer sciences, Peking may have known more about its offshore oil potential for much longer than has been generally assumed in the West. Reliable Japanese sources indicate that by 1965 Peking had conducted an aeromagnetic survey of the Po Hai Gulf with French technical help, as well as several additional surveys of portions of the gulf and the Yellow Sea by ship on their own, utilizing the relatively elementary "air gun" method. A Japanese visitor to China in 1965 found intense interest in offshore drilling techniques among students at the Peking Oil Institute and was shown an exhibition of elaborate handmade models of the latest drilling rigs then in use throughout the world.[7] Since then, Peking has gradually accelerated its offshore activities, selectively importing foreign technology but taking great care not to let this process get too far ahead of the development of Chinese expertise. Step by step, Peking has upgraded its technical capabilities, while simultaneously proceeding with a domestic rig-manufacturing program of its own in which its new imported rigs and components have been used, in effect, as working models.

Operating at first from fixed platforms in the shallowest water, Chinese oil technicians then contrived their own jerry-built, barge-style rigs, drawing on Romanian, Japanese, and other imported components. Significantly, Peking appeared to recognize the deficiencies in Romanian offshore technology at a relatively early stage and made cautious secret overtures to Japan from the very outset of its offshore activities. Thus, in the summer of 1966 Peking sent a high-level technical mission to Japan that was "kept secret from the press at the time" in order to avoid criticism from the United States.[8] The mission inspected a variety of the Japanese offshore drilling rigs then in use or under construction and obtained all or part of the technology for a barge-style drilling rig that was subsequently built in China in 1967, the *Pinhai I*.[9] In the hierarchy of offshore drilling technology, the barge-style rig is the easiest to construct but the most limited in its uses and is surpassed, in ascending order of technical complexity, by the drillship, the jack-up, the catamaran drillship, and finally, the semisubmersible, which has floats that can be submerged and thus provide greater stability and maneuverability in rough seas than other types of rigs.

Japanese sources state that the *Pinhai I* bore a "close resemblance" to one of the more rudimentary barge-type models then in use by a Japanese firm, the Asia Offshore Drilling Company, and strongly imply that the blueprints were made available at a token price in the hope of getting an inside track on future rig sales. Capable of operating in water 100 feet deep and of drilling up to 1.2 miles, the *Pinhai I* was not readily movable from place to place but was adequate for the shallow-water operations then getting under way in the Po Hai Gulf. In mid-1968 the Chinese began test drilling operations in the southern portion of the gulf with their new rig and made their first strike in the fall of that year. This led to the test production of small submarine pipelines.

By 1969 Peking was ready to move to the next step on the technological ladder and ordered an $11-million drillship, dubbed the *Pailung*, from Ishikawajima-Harima Heavy Industries. Like the earlier purchase of barge-style rig technology, the fact that the *Pailung* had been purchased from Japan was not to become known to the rest of the world until several years later. Negotiations were also initiated in 1969 for another, more sophisticated drillship incorporating U.S.-licensed components, but the deal was called off in April 1970, when Peking laid down its "Four Conditions" barring dealings with firms then engaged in trade with Taiwan. In 1970 Peking improved upon the *Pinhai I* with the *Pinhai II*, a new model of barge-type rig, boasting a self-elevating deck. In this case, as in others, it is not clear how much of the rig was Chinese in manufacture and to what extent components were imported. In any case, Peking was confident enough of its increasing capabilities by 1970 to initiate negotiations with Mitsubishi for a used jack-up rig, the *Fuji*. The negotiations were ultimately concluded in late 1972 after agreement on a price of $9 million. The *Fuji* was delivered in early 1973 and was to become the prototype for initial Chinese manufacturing efforts in this crucial area of rig production. Aided by intermittent visits by Japanese technicians over an initial ten-month trial period and by a 1972 mission of Chinese technicians to an offshore technology conference in Tokyo, the Chinese attempted to proceed on their own, drilling a dry hole in their first venture with the *Fuji* fifteen miles east of

Tientsin on 24 April 1973. Four or five additional wells each produced modest discoveries, and by May 1974, according to Japanese sources, Peking called back the Japanese engineers to help with repairs for the last time. The rig was then moved to a point forty-six miles offshore, where an oil field extending over a twelve-square-mile area was soon discovered.

As a result of the Chinese preoccupation with secrecy and the desire of foreign firms dealing with the Chinese to avoid alienating Peking, there has been some confusion surrounding the precise number of offshore rigs bought by the Chinese prior to 1974. Just as the Chinese acquisitions of the *Pinhai I*, the *Pailung*, and the *Fuji* were all handled on a hush-hush basis, so there is reason to believe that one or more additional purchases occurred during this period. In particular, it would appear that China imported at least one other foreign jack-up in addition to the *Fuji*, containing more sophisticated components than the *Fuji*, apparently for use as an experimental model rather than for immediate operation. Authoritative American sources indicate that American and Japanese business cooperation led to the covert sale to China of a jack-up following the 1972 Nixon visit to Peking. As in the earlier Japanese transactions, this deal was kept quiet at the insistence of Peking, not only because the Chinese relationship with Tokyo and Washington was still in a tenuous stage, but also because the "self-reliance" mystique ruled out the public acknowledgment of foreign technological purchases even from Communist sources.[10] The only public hint of the transaction came in early 1973, when the president of Asia Offshore Drilling visited Peking and was reported to have completed negotiations for the sale of a jack-up.[11] In early 1975 an official of the powerful Japanese construction firm Ohbayashi-gumi, returning from a mission to China, wrote in the journal of the Japanese business organization, Keidanren, that China then had two jack-up rigs.[12]

The precise number and character of the rigs that have been obtained by China is of interest only insofar as it helps to gauge the progress of a "self-reliance" effort in which each rig imported provides new elements of technology useful in

China's own rig-manufacturing program. Regardless of whether one or two foreign jack-ups were imported prior to 1974, what is important for our purposes is that Peking had already decided at that point not to import foreign equipment or technicians on a wholesale basis for its offshore development. The decision to purchase the *Fuji,* a secondhand rig, served clear notice that Peking intended to rely primarily on a do-it-yourself effort, while continuing to acquire progressively more sophisticated prototypes for experimental purposes as the process of trial and error proceeded. The purchase of the *Fuji* came in the aftermath of the Nixon visit, at a time when the Western majors were eyeing the possibility of collaboration with China, and was hailed by pro-Peking Chinese observers in Hong Kong as "an indication that China will rely on its own strength to develop its undersea petroleum." [13]

The fact that Chinese imports were seen as part of a "self-reliance" approach rather than a drive for rapid production as such was apparent in the Chinese refusal to engage in lease arrangements in lieu of rig purchases following the opening of contacts with the United States in 1972. Many Western and Japanese rig companies were quick to recognize that the Chinese wanted to make their own rigs. These companies concluded that it would be more profitable to enter into a series of lease arrangements than to sell a limited number of prototype rigs, unless sales could be made at high enough prices. Despite persistent overtures for lease arrangements on attractive terms, however, the Chinese decided to bide their time and watch for variable shifts in the world rig market.

Self-reliance
and Offshore Drilling

The delivery of the *Fuji* in early 1973 was followed by a triumphant Chinese announcement in December 1974 that China had completed its first domestically made jack-up, the *Po Hai I*, and had drilled its first test well with the new rig. This prompted speculation in American oil journals [14] that the rig was really the *Fuji* or that the Chinese had copied the Japanese model in whole or in part. According to Peking, however, the two events were completely unrelated. In its first version of the *Po Hai I*'s history, issued at the time of its 1974 announcement, China said that the rig had been "put to the test for two years" prior to that time. [15] Two years later Peking altered the record slightly, pointing to the *Po Hai I* as a casualty of Lin Piao's "revisionism." As early as 1961 there was an attempt to design a Chinese-made jack-up, Peking said, but this had been frustrated by "some authorities" who wanted to buy or lease rigs instead, touching off a "fierce struggle between the proletarian and bourgeois lines" that paralyzed action. In 1966 a second design of the *Po Hai I* had been completed and approved under the impetus of the Cultural Revolution, only to be shelved again as a result of "revisionist" delaying tactics. Finally, this account concluded, construction of the rig did start in 1970, resulting in its first test well in May 1971. [16]

While it is not possible to judge with finality what actually happened, my investigation suggests that the case of the *Po Hai I* exemplifies the way in which China's "self-reliance" approach is consciously married with the selective import of foreign prototypes and components. Negotiations for the *Fuji* had started in early 1970, and the blueprints were reportedly sold at a token price as a come-on well before the delivery of the rig itself in 1973. China modified the design, however, and had to get Japanese help in untangling technical snarls. During a visit to Tokyo in November 1974 I was told by a variety of Japanese sources in close touch with China that full-scale construction work on the *Po Hai I* had started in January 1973 at the Hongchi shipyard at Talien and that Japanese engineers had been consulted on a number of points related to the construction of the *Fuji* as well as its operation when the rig was delivered in April of that year. After the completion of the *Po Hai I*, the *Fuji* was renamed the

Po Hai II. Two smaller, shallow-water jack-ups, capable of drilling in water 75 feet deep, were also completed in 1973.

Chinese descriptions of the *Po Hai I* have not gone into much technical detail and have come primarily in the context of popular articles focusing on the rig as a triumph of "self-reliance." In its initial news announcement Peking said that the rig had a 5,000-ton hull resting on four legs, each measuring 250 feet in length and 8 feet in diameter.[17] A more detailed subsequent discussion added that the rig was a self-escalating but not self-propelled rig capable of drilling in waters 190 to 220 feet deep and had been operating in the area between Talien and Tientsin. Hailing this "magnificent structure," the article declared that when the drilling rig is being "towed away," the body of the rig "floats on the surface of the water with its four giant legs standing erect on the deck, and together with the derrick, they form a beautiful and magnificent picture."[18] At the height of the factional controversy following the ouster of Deputy Premier Teng Hsiaoping, a 1976 account spotlighted the *Po Hai I* as a prime example of how "self-reliance" can be made to work if all of China's technical know-how is brought to bear on the task at hand. Thus, the article said, bridge engineers helped to design legs strong enough to hold up the rig.[19]

Analyzing the *Po Hai I* on the basis of the limited information available, a leading American drilling company executive, John A. Sage, vice-president of Fluor Drilling Services, said that it is "substantially similar" to many American jack-up units operating in the Gulf of Mexico. Sage indicated that the *Po Hai I* would be well suited to the Po Hai Gulf, the Yellow Sea, and the relatively close-in coastal areas of the East China Sea but does not have legs long enough or strong enough to withstand the tumultuous wind and waves characteristic of the East China Sea during most of the year. Sage claimed that with legs 220 feet long, a jack-up of this type could operate only in water up to 180 feet deep and only outside the typhoon and monsoon seasons.[20]

On the basis of the experience gained in operating and

manufacturing jack-ups through the purchase of the *Fuji* and the construction of the *Po Hai I*, the Chinese felt emboldened to take the next step in upgrading their rig technology by placing orders in 1974 for two more advanced jack-ups from Robin Loh,[21] a Singapore shipyard with Japanese and U.S. links. The two Robin Loh rigs, *Nanhai* ("South Seas") *I* and *II*, were put into service in early 1977. In contrast to the 180-foot depth capability of the *Po Hai I*, they are capable of operating in 275 feet of water. Moreover, they contain a variety of relatively up-to-date U.S. components, including Armco pumps and hydraulic equipment. One of the key conditions of the contract was that Chinese technicians could observe the construction work, and from seven to ten Chinese were intermittently on the scene in Singapore beginning in January 1976. This would appear to indicate clearly that Peking hopes to incorporate as much of the Robin Loh technology as possible into its own manufacturing program. The initial contract did not clearly specify how much technical assistance would be provided following delivery of the rigs, and most indications were that Peking did not ask for American help in operating them. J. Ray Pace, president of the Baker Trading Company, who has had intensive contact with Chinese oil officials since 1972, praises Chinese technical skills but questions whether it will be possible to learn the ins and outs of the Robin Loh equipment without some assistance. "They've taken a mouthful they can chew," Pace observed, "but just barely."

In addition to their progress in learning how to manufacture jack-ups, the Chinese have also succeeded in making a rudimentary catamaran drillship, *Kantan* ("Explorer") *I*. As "floaters," catamarans and semisubmersibles are more maneuverable than jack-ups and thus involve correspondingly greater technical complexity. Almost simultaneously with its announcement that the *Po Hai I* had drilled its first well at an unidentified location, Peking disclosed in late 1974 that *Kantan I* had "successfully drilled an oil well in the southern part of the Yellow Sea."[22] Emphasizing that the rig was wholly domestic in manufacture, Chinese accounts said that it had been constructed in Shanghai by joining two cargo ship hulls side by side. Peking expansively reported that the *Kantan I* had "withstood 30 mile per hour gales and gusts of over 40 miles per hour without hull damage" and had operated in "fairly deep waters." However, skeptical Western oilmen speculated that the test drilling might have been conducted in a cove protected from strong winds, contending that "floaters" require sophisticated dynamic positioning and tension-maintaining devices to operate effectively in deep waters and that Peking does not yet have this technology. As if in response, Peking later declared that *Kantan I* had "successfully carried out its first winter drilling test in 1975 under complex weather conditions" and had accumulated 160 days at sea during the year. But no mention was made of where the rig had operated or the water depths involved.[23]

Most indications suggest that *Kantan I* was made without reliance on an imported prototype. Western rig experts, shown a photograph of the rig, said that it was a simplified adaptation of a well-known type of catamaran long operated by an American company and that the general design for such a rig would not have been difficult to obtain.[24] By contrast, in attempting to move to semisubmersibles, Peking has confronted major technical obstacles and has long recognized that it would be necessary to acquire at least one imported prototype. However, the desire to purchase one or more semisubmersibles has proved to be more controversial than other rig imports in domestic political terms because the large foreign exchange outlays involved provide a more conspicuous target for advocates of undiluted "self-reliance."

As early as December 1973 Peking had signed a $22.6-million contract with Mitsubishi for the semisubmersible *Hakuryu* ("White Dragon") *II*, only to back out at the last minute after an unsuccessful effort to get the price lowered.[25] At about the same time China had initiated inquiries regarding the Aker H-3, a more advanced Norwegian semisubmersible with a capability of operating in 1,000 feet of water, in contrast to 600 feet in the case of the Japanese rig. The H-3 was to have been built by a Singapore shipyard, Far East–Levingstone, and preliminary agreement was reached on a price of $40 million per rig. By April 1974, however, when a team of ten Akers technicians arrived in China to finalize the deal, the rig market had gone up dramatically and the price for the H-3 had doubled. Percival Meadows, board chairman of Far East–Levingstone, has blamed the price increase for the Chinese decision to hold off on ordering the H-3. Unfamiliar as Chinese oil officials were then with the ups and downs of the world rig market, Meadows said, the doubling of the price within a year was unfathomable in their eyes, and they were even more astonished to find that the price had gone up still further when they sought to reopen negotiations in September 1974. At that point, Peking abruptly shifted its search for semisubmersibles from Singapore to Western Europe, the United States, and Japan, hoping to capitalize on the possible emergence of a surplus in the world rig market.

Akers and other Norwegian rig-makers were approached for rigs that were already in the manufacturing pipeline, leading to a continuing exchange of Norwegian and Chinese missions during 1975 and 1976, marked by exhaustive Chinese inspection visits to rig construction sites but no firm orders. In tandem with its exploratory Norwegian exchanges, Peking entered into formal negotiations with an Italian firm, Saipem, for the construction of five Scarabeo III semisubmersibles by the Blohm and Voss shipyard of Hamburg. A preliminary agreement had been initialed in April 1975, but the deal was still in abeyance in mid-1977, and the number of rigs contemplated in the agreement would not be more than one or two, Chinese sources said, even in the event that the transaction was actually consummated. Equipped with U.S. components, the Scarabeo III is a self-propelled, self-positioning, three-legged semisubmersible with a depth capability of 1,200 feet that apparently attracted Chinese interest as a rig peculiarly suited to the stormy waters of the East China and Yellow seas.[26] However, it is understood that the steep $350-million price tag of the original deal for five rigs became a political issue in the Left-Right factional struggle, and Peking was also actively investigating semisubmersibles in Poland, Japan, and the United States during 1975 and 1976 while the Scarabeo deal was left dangling.

In the rig trade, there are some cynics who accuse the Chinese of a canny variety of "window shopping" that borders on patent theft. Peking's "negotiations," it is alleged, are mainly a way of getting a feel for the latest foreign technology and of gathering ideas for use in Chinese rig-making. This suspicion is widespread and gains in intensity as the Chinese prolong negotiations such as the H-3 and Blohm-Voss transactions, but it is not shared by many of those who have dealt most intensively with the Peking agencies involved. The Chinese are afraid of being bilked, it is said, and are cautiously studying the ups and downs of the world rig market. Percival Meadows, who conducted the 1974 negotiations on the H-3 with the Chinese, dismisses the window-shopping charge as "absurd." Meadows had told newsmen after his return from China: "Our meetings would start at 8:30 and carry on straight into the evening. Sometimes as many as 30 experts at a time would be there to ask questions." [27] Asked in an interview if this did not suggest an attempt to acquire a close knowledge of the H-3 without necessarily intending to buy the rig, Meadows told me that the questioning was "systematic" but was focused on the precise operational methods and capabilities of the rig rather than on the specifications that would be relevant for manufacturing. Meadows stressed the rapid increase in price that had occurred between the original Chinese overtures for the H-3 and the critical stage of negotiations nearly a year later.

In a limited sense, there may be some truth in the window-shopping complaint in that Peking may intend to accompany its purchase of one or more prototypes with the subsequent import of components on a selective basis. By familiarizing themselves in detail with different rig models, Chinese technicians can differentiate between those components that can be copied or adapted without detailed specifications and those that must be either imported or, possibly, manufactured in China under licensing arrangements. This would then permit the import of prototypes to be integrated more effectively with domestic manufacturing plans. Chinese officials frankly told a visiting American in 1976 that technicians at a Shanghai shipyard were studying the extent to which semisubmersibles could be manufactured in China before deciding on foreign orders.

It should be emphasized at this juncture that Peking has already relied on imported components in filling key technological gaps in its rig-manufacturing program. In addition to the high-grade alloy steels needed to withstand high pressures and temperatures (see chapter 2), Peking may well be importing mud circulators, drilling tools, and the pneumatic engines that make offshore rig platforms go up and down. China is known to import as many as thirty drilling rigs from Romania annually, some in barter deals as payment for crude, and these could be used in offshore as well as onshore rigs.

One interpretation often offered in place of the window-shopping allegation is that Peking has been serious in its negotiations but has been deeply suspicious of the way in which Western rig companies seek to structure their contracts. Unlike Far East–Levingstone, most of the Singapore firms approached by China in 1974 were not willing to make even a tentative price commitment in advance. Instead, they insisted on their standard practice of pay-as-you-go arrangements in which the purchaser pays for the rig in installments, during its manufacture, rather than in a single payment on final delivery. This practice enables the company to avoid tying up all of its money during the protracted periods required to manufacture a complete rig and to retain some flexibility as a hedge against fluctuations in the prices of key components. Often, Western rig manufacturers have also sought to make the sale of rigs conditional on follow-up arrangements for the participation of their technical personnel in the operation of the newly purchased rig, not only as a means of ensuring its successful use, but also to keep technicians busy during periods of shifting demand.

In the view of some observers, it has been the Chinese refusal to accept such conditions that has often led to the breakdown of negotiations. In any case, even if the window-shopping thesis should prove valid, this would be of greater significance to rig companies than to U.S. policymakers. What matters in political terms is that China appears likely to have a substantially augmented fleet of rigs in the foreseeable future and will, accordingly, be able to conduct its own test drilling in contested areas where others have hitherto had a clear field. If Peking does buy rigs on the world market in quantity, this would speed things up; but in any case, rig purchases abroad to some extent would appear to be necessary and would go hand in hand with accelerated domestic rig construction based on foreign prototypes.

The degree of urgency with which Peking approaches the acquisition and manufacture of semisubmersibles and other deep-water drilling rigs will be of particular importance in assessing future Chinese strategy with respect to contested areas of the continental shelf. Semisubmersibles and certain varieties of drillships are necessary for oil exploration in many parts of the continental shelf where the wave action is too strong for other rigs or the water is too deep. For example, in the East China Sea, even though the average depth of the continental shelf does not exceed 600 feet and wide areas are relatively shallow, the shelf also has numerous pockets of deep water where a jack-up rig could not operate. A go-slow approach to semisubmersibles would not necessarily rule out Chinese efforts to explore in contested areas with jack-ups, but it would be more likely to accompany a cautious Chinese posture. By the same token, the acquisition of semisubmersibles would not assure aggressive Chinese tactics but would strongly suggest that Peking is ready to press its claims in contested areas.

Among Western oilmen who deal intensively with Peking, there is little doubt that the Chinese will move steadily into deeper waters and will be able to manufacture their own semisubmersibles in a relatively short span of years. Lawton Laurence, a veteran American oil-equipment manufacturer in Singapore, declares that "ten years from now, they'll be where we are today, semisubmersibles and all. There is no way that they can stay up to date by doing it themselves. They will always be behind. But they can still have a major industry." Reflecting on his negotiations with Chinese officials, Percival Meadows of Far East-Levingstone observed that Peking "is certain to explore in the deeper waters of the shelf because the bonanza area is probably in the deeper water, and they know it." During his H-3 negotiations, Meadows recalled, the Chinese repeatedly asked whether the Akers rig "could actually drill effectively to a 1,000-foot depth as we said. They were very concerned about this point."

Even though the technology used in semisubmersibles is more complex than that used in other rigs, Meadows declared, it is not beyond the Chinese,

and we know because we make the bloody rigs. Drilling companies are selling their services, and naturally, they're going to make it sound as complicated as possible. But I have no doubt that the Chinese will be able to learn to use a semisubmersible rather quickly. This will be similar to the *Fuji* pattern. There will be instruction but not continuing assistance in management and operation. After trial runs for two to three years, they will try test drilling in deeper waters. By that time, they will also be far along in trying to manufacture some form of semisubmersible. This is no longer a monopoly of Texas and Louisiana, you know. Texans aren't the only ones who can drill for oil in the ocean. If the Chinese can manufacture and operate a submarine, they can acquire the know-how for a semisubmersible. It's all a matter of determination and priorities. You Americans are ready to acknowledge that people like the Norwegians can do it, so why can't the Chinese?

Self-reliance
and Offshore Surveys

As in the case of drilling rigs, Peking has sought to minimize foreign dependence in developing an offshore seismic survey capability of its own by combining carefully modulated foreign imports with the parallel development of Chinese operational and manufacturing expertise. Since seismic and other offshore survey work involves even more arcane areas of specialized technology than rig-building, however, Peking has found it more difficult to do without foreign help.

As we have noted earlier, most evidence suggests that the Chinese fleet of some thirty-five oceanographic research vessels of various kinds has included at least three ships with a marine geological capability for at least a decade. A Japanese oceanographic mission to China in 1975 reported that one of these, the *Tung Fang Hung* ("Red East"), had been in service since 1954 and that another shown to the group had reportedly been built in 1967.[28] The initial geophysical survey of the Po Hai Gulf in 1966 was conducted by Chinese vessels utilizing a Toran-style navigation system purchased in France and the relatively primitive "air gun" survey method. Soon thereafter Peking acquired a widely used Western type of onshore seismic survey equipment from a subsidiary of the French geophysical company Compagnie Général de Géophysique (CGG) and proceeded to adapt this equipment successfully for marine use. Although it is not known precisely when Peking deployed its first Chinese-made seismic survey boats, a detailed model of one was on display in the Shanghai Industrial Exhibition in early 1973. When a visiting American business executive, Raymond Cox, president of the Geospace

Corporation of Houston, asked how many of these China then had in operation, he was told that two of them had been successfully used "for some time." [29]

Beginning in 1972, as part of the general softening in the implementation of its "self-reliance" policy that followed its new contacts with Washington and Tokyo, Peking sought foreign help more directly for its onshore and offshore oil survey programs. In August 1972 China bought $120,000 worth of sophisticated Canadian geophysical equipment at a Canadian trade fair in Peking, a small order by Canadian standards but a large one for Peking and extremely significant as a technological stimulus in that it consisted of prototypes of a variety of key items from digitizers and computer-equipped spectrometers to magnetometers,[30] all of which are used in offshore as well as onshore oil survey work. In September 1973 the Continental Oil Company (Conoco), hoping to pave the way for crude oil purchases, sold Peking an onshore seismic survey processing system known as Vibroseis, complete with a training program for Chinese technicians. Finally, in early 1974 Peking began to shop actively for completely equipped seismic survey vessels that would enable them to keep abreast of the latest offshore technology.

In an unusual departure from the "self-reliance" policy, Peking decided to get a concrete example of what foreign technology had to offer by engaging a CGG seismic ship, the *Lady Isabel,* to carry out a 1,240-mile offshore survey under a special contractual arrangement in which Chinese crews worked side by side with the boat's French crew. At the same time, in April 1974, the Canadian Department of Energy, Mines, and Resources was invited to send a geophysical mission to China, and it soon developed that Peking wanted to purchase the necessary equipment for three seismic survey boats.[31] One of the firms represented on the mission held detailed discussions with Chinese officials, but Peking eventually bought the *Lady Isabel* from CGG. The French boat was delivered in the summer of 1974.[32] In addition, Peking also bought two 500-ton, multipurpose oceanographic vessels from the Japanese trading combine Sumitomo, relying partly on Chinese-made equipment in outfitting these boats and partly on selective imports. The *Lady Isabel* was renamed the *Liao Yuan* ("Prairie Fire"), and the other seismic survey boats have since been identified as the *Hsingho* ("Spark") and the *Fentou* ("Struggle"). The two new multipurpose vessels, primarily devoted to geological work, were named the *Haiyang* ("Ocean") *I* and the *Haiyang II*.[33]

   Peking's relations with Canadian geophysical companies have illustrated the manner in which China maximizes the use of its Western and Japanese commercial contacts in developing its own manufacturing program. In connection with their participation in the Canadian Electronic and Scientific Instruments Exhibition in 1974, Canadian geophysicists were invited to lead seventeen specialized seminars that lasted for five days; Peter Hood, leader of the Canadian group, recalled that the businessman who had unsuccessfully attempted to sell a survey boat was subsequently "kept in Peking for many months having periodic meetings, where they got him to supply technical information (this included a number of seminars) and costs of equipment bit by bit." Hood praised the courtesy of the Chinese and their readiness to exchange information in the field of basic geophysical research but concluded that a worthwhile exchange of information in exploration geophysics is not "presently possible, since the vast amount of information imparted was in one direction; that is, from the Canadian to the Chinese side." When a group of Chinese geophysicists paid a return visit to Canada in November 1975 the mission returned with a 129-pound shipment of technical brochures provided by Canadian companies.[34]

   The most serious problems confronted by Peking in developing an offshore geophysical capability have not related to the purchase of survey vessels as such, but rather to the acquisition of certain equipment for these boats that cannot be sold by Japanese and Western companies without the approval of the interallied screening committee known as

COCOM (Coordinating Committee for the Control of Strategic Trade with Communist Countries). In the case of American companies, some items are directly covered by export-control regulations that extend to foreign subsidiaries or licensees. The general U.S. policy guideline governing export licenses in the case of oil-related equipment has been that the development of energy resources anywhere will make it easier for the United States to meet its own energy needs and that the U.S. government should neither actively encourage nor actively discourage sales of such equipment. Thus, the United States approved the sale of Magnavox satellite navigation equipment for the CGG boat, since the same equipment had previously been made available to the USSR. Certain other types of equipment needed in oil exploration also have a direct military applicability, however, and the United States has blocked sales in these cases. As one example, Peking has unsuccessfully attempted to obtain certain types of gravity meters that could have a military application in measuring the gravitational environment of missile-launching pads. Another area of difficulty for the Chinese has come in the acquisition of certain U.S.-licensed magnetometers with unusually high precision and corrosion resistance. Magnetometers are crucial not only in oil exploration but also in submarine detection, and China wanted to buy a highly sophisticated variety of nuclear precession magnetometer known as the pulsed-rubidium-vapor type, which utilizes technology related to the latest U.S. laser–maser technology. In this case, too, the United States blocked sales to China, although it is not clear how effective the embargo has been in view of the fact that the patents covering the main elements of the technology involved have been registered since 1960 and could have been pirated by the Chinese.

Peter Hood, who is director of geophysics and geochemistry in the Canadian Geological Survey, reported after the 1974 Canadian geophysical mission to China, that "the Chinese have quite a good capability for the manufacture of geophysical equipment, probably better than that possessed by the U.S.S.R." [35] During their meetings with the Chinese, the Canadians found that "the questions which were posed were usually quite pertinent and indicated that the questioners had a fairly advanced knowledge of the topic at hand." [36] In particular, the mission noted that nuclear precession magnetometers were being manufactured in a Peking factory and that the cesium and rubidium units needed for the pulsed-rubidium-vapor type were visible in the factory. The mission was also shown three magnetometers of the simpler proton precession type, also commonly used in offshore survey work.[37] The level of sensitivity in magnetometers banned under a COCOM regulation [38] is a magnetic intensity of one gamma or higher, and the Canadian group found a proton precession magnetometer said to be of this sensitivity in the Peking factory. However, it was not clear whether the nuclear precession magnetometers were up to Western standards either in the precision of their sensor elements and timing circuits or in the degree of their corrosion resistance. The gravity meters seen in the factory were judged to be "not of high quality," which would explain Chinese efforts to buy prototypes of advanced types abroad. Similarly, Peking has sought to reinforce its domestic manufacturing efforts with such magnetometer sales as Canadian firms would make following the 1974 mission, as well as with purchases from the United States.[39] When a group of leading American seismologists visited China in November 1974, the magnetometer laboratory was more carefully shrouded than in the case of the Canadians, and the mission reported only that work on a pulsed-rubidium-vapor magnetometer is "in progress." [40]

Reviewing its achievements in offshore survey work in 1976, Peking pointedly mentioned a nuclear precession magnetometer among the seismic instruments then in use, observing that it was "up to advanced standards" and further stating:

Chinese surveyors who could do only seismic prospecting and a few other individual tests before the Cultural Revolution are now able to carry out multi-purpose surveys combining seismic, magnetic, gravimetric and sounding tests; stratigraphic section, water sampling, and solid mineral tests; and submarine sampling. They are not only capable of making general but also reconnaissance and detailed surveys to provide information on well locations.[41]

Most seismologists who have visited China believe that Peking will have relatively little difficulty in catching up with the West in the geophysical instrumentation needed for oil exploration and will not have to rely much longer on imported equipment. Initially, the 1974 U.S. mission reported, Chinese scientists have focused their geophysical work on earthquake prediction, but their technical competence is more than equal to the needs of an oil-oriented program.[42] A more difficult bottleneck for China appears to lie in the computer technology needed for the interpretation of seismic data. It is the sophistication of the data processing utilized in interpreting survey findings that determines not only how quickly the processing is done but also how precisely. To a great extent the technological lead enjoyed by American and other Western oil companies over new entrants to the field lies in the use of large computers with a capacity for "microprogrammability" that enables them to store and recall a bewildering variety of data rapdily. This is an especially significant factor in offshore survey work, since seismic data are generated more rapidly by boats than by more cumbersome onshore survey methods. On their own, the Chinese have already gone far in computer technology, but not in the development of the large computers most appropriate for large-scale offshore surveys or the type of intricate programming "software" that has evolved through trial and error in Western geophysical work.

Peking frankly recognized its limitations when it concluded its 1974 agreement with CGG for survey boats and ordered two Control Data Cyber 172 computers as part of a $7-million package deal that included programming tailored for these computers.[43] The Cyber 172 is one of the larger U.S.

computers, however, and has a main frame, or "memory" unit, that could be adapted for use in antisubmarine warfare or air-defense systems. Under U.S. export-control regulations, the computer could not be sold to China without an "end use" agreement providing for some form of inspection or other safeguards. In November 1976, nearly two years after the original agreement between Peking and CGG, Peking and Washington finally found a formula that permitted the sale of the computers,[44] but new difficulties soon arose in 1977 over a Chinese bid for four more of an equally advanced model. In any case, the Chinese have been operating three seismic survey boats since late 1974 and have been laboriously processing the resultant tapes with other equipment in their computer center in Peking. China has a relatively advanced computer of its own, known as the Model 111,[45] and has also been importing other American computer technology less advanced than that exemplified in the Cyber 172. A $5.3-million deal with the Geospace Corporation of Houston in late 1973 provided for a Raytheon 703 computer and related American equipment for use in seismic processing work. Twenty Chinese were trained at Geospace headquarters in Houston under this arrangement in late 1974, and 70,000 pounds of seismic processing equipment was airlifted to Peking on 25 January 1975 in the first commercial jet cargo flight between the United States and China. Consisting of ten field systems and three office playback centers, the Geospace equipment was intended primarily for onshore seismic studies, but Raymond Cox, president of Geospace, said that offshore tapes could be adapted for processing with this equipment, although it would take longer than with the use of larger computers.

In the computer field, as in the seismological sciences, Peking has laid the foundations for relatively rapid adaptation to offshore seismic work. Raphael Tsu, the leader of a group of U.S. computer experts who visited China in 1972, reported that "the technical competence and (to a lesser degree) the hardware I saw in many cases approached what we are accustomed to seeing in the U.S." Tsu said that Chinese computer engineers were engaged at that time in

substantial programs on large-scale integration and are encountering the same kind of problems we faced here a few years ago. For example, the relatively poor quality of some of their materials makes it difficult to achieve high enough yields to build sophisticated memory units with this novel technology. They seem, however, to be establishing a foundation that will in a few years enable them to master the art of large-scale integration.[46]

Significantly, by May 1975 China had announced its "large, hybrid electronic analogue computer," Model HMJ-200, said to combine the advantages of analogue and digital computers.[47] Another highly favorable estimate of Chinese capabilities in computer technology came from a group led by Thomas E. Cheatham, Jr., director of the Harvard Computer Research Center. The Cheatham report stated:

While China is clearly not the equal of the United States in computing at present, evaluations such as, "China is N years behind the United States" are, we feel, most hazardous. The Chinese have demonstrated the ability to reach very high levels of technology in a very short time.[48]

At the same time, the report pointed to bottlenecks in connector technology and in the orientation of Chinese computer technology to numerical computation for scientific and engineering calculations rather than to the type of data processing needed in handling geophysical data. Model 111, the intermediate-sized computer believed to be in use at the Peking processing center where offshore seismic data are being processed, was described as "well designed but lacking in complex input–output equipment as a reflection of the Chinese emphasis until now on numerical computation."

Noting that the Model 111 had integrated circuits, the Cheatham report pointed out that the Chinese had been manufacturing integrated circuits at least in limited quantities since 1968 and had made "impressive" strides in learning how to make the high-performance variety.[49] This observation was of special interest in view of a subsequent statement by visiting Chinese computer experts that the first integrated circuit computers in China had been specifically designed for oil surveys.[50]

The Future:
How Fast and How Far?

Reviewing the state of Chinese offshore capabilities as of 1977 and looking ahead to future prospects, it is necessary to distinguish between the shallow waters of the Po Hai Gulf and other close-in areas, where Peking is on the threshold of significant production, and the deeper waters of the continental shelf, where Peking faces a more serious technological challenge.

As we have seen, Peking began test drilling in the Po Hai Gulf in mid-1968 and has made at least one known discovery of a structure extending over a twelve-square-mile area. Chinese sources have continued to evince enthusiasm regarding the potential of the gulf, but specific information as to the number of wells drilled and the discoveries made has been closely guarded. A New China News Agency report on Chinese crude production in September 1976 said only that "a number of high-yield wells have been sunk in the Gulf of Po Hai." A Norwegian oilman who has had business dealings with Chinese officials states that twenty exploratory wells had been drilled in the gulf by 1976, apart from the wells drilled in other coastal areas.[51] This is a plausible but perhaps low figure, considering the definitely known offshore rig capabilities discussed earlier (two barge-type rigs, one drillship, six jack-ups, and one catamaran drillship). Except for the catamaran and one of the two Singapore jack-ups, sent initially to the South China Sea, the other eight rigs are all known to have been deployed primarily or exclusively in the gulf. In any case, exploration in the gulf will escalate rapidly following the completion of four more Chinese-made jack-ups similar to the *Po Hai I* and two more catamarans similar to the *Kantan I* that were under construction in mid-1977, although all of these rigs will also have the capability for drilling in the Yellow Sea and the less stormy parts of the East China Sea. At least two of the new jack-ups, it should be noted, are slated for assignment to the Shantung government, which has jurisdiction in the Yellow Sea and the northern East China Sea, as distinct from the Tientsin government, which controls oil operations in the Po Hai Gulf.[52] Six new docks constructed near Shanghai in 1976 were linked partly with the growing need to accommodate offshore drilling vessels.[53]

According to Japanese sources, by 1976 only one production platform requiring the use of offshore production technology was in operation in the Po Hai Gulf. Located forty-six miles east of Tientsin at the site of the 1974 strike cited earlier, this appears to be the same one cited by a CIA analyst.[54] It is not to be confused with earlier reports of "several wells" in the estuary of the Yellow River and of "floating docks" in close-in offshore areas near Tsingtao,[55] which involve shallow-water operations from dikes and utilize simple adaptations of onshore techniques. At the same time, judging from the nature of certain orders for offshore support equipment known to have been placed by China and the timetables for delivery, it would appear that Peking had not only embarked on accelerated exploration activity in the Po Hai area by 1976 but was also ready for the expansion of its offshore production program. Thus, two sophisticated diving-chamber complexes delivered in 1975 by a Sumitomo subsidiary can function at depths of more than 300 feet and can be used not only for positioning rigs but for laying undersea cable and for welding pipes under water.[56] Similarly, other types of oceanographic equipment delivered by Japanese companies in 1975, such as echo sounders, side-scan sonars, marine safety equipment, and computer-controlled depth-measuring systems, also have a dual utility for production as well as test drilling operations.

In the case of work ships or supply boats, which can service drilling rigs and production platforms alike, the number ordered would clearly appear to suggest that extensive production operations are either under way or are imminent. Two or three supply boats are generally adequate to service one rig. In addition to the work ship *Kuroshio,* ordered in conjunction with the *Fuji* from Japan, Peking has purchased sixteen oil-rig supply tugboats from a Danish firm, eight of which had been delivered by late 1976, plus another five from Japan. In both cases, the ships are suitable for transporting men, equipment, and supplies from onshore terminals to offshore drilling stations and are equipped with large afterdecks for transporting used materials back to shore, as well as with mooring devices designed primarily for production operations. Most Western oil companies use helicopters to supply their drilling rigs, and in 1976, Peking placed a $2.3-million order for four Bo-105 Messerschmitt helicopters, while continuing negotiations for an additional fifteen. With its two turboprop engines, the Bo-105 is regarded as ideal for supplying offshore rigs as well as for maritime military uses.

The most revealing indication that Peking has made significant strikes and is approaching major production has come in an order for three well-logging units specifically suited for offshore use as part of a larger order including ten more for onshore use. Well-logging is the critical step in making a precise survey of the geological formations in a well after a discovery is made but before production casing is actually installed. China makes its own well-logging units but purchased a highly sophisticated type with a late-model minicomputer from Dresser Industries of Houston for delivery in April 1977.[57] Each of the three cumbersome units would normally be used with one rig but could be transported from one to another, if necessary, and thus utilized with more than three rigs. Another significant indicator of preparations for offshore production has been the acquisition of large-diameter, corrosion-resistant pipeline for undersea use from the offshore construction affiliate of ENI, the Italian state oil enterprise, and from Nippon Kokan of Japan. As of 1976, these purchases were in relatively small quantities, and it was not clear whether China was attempting to make its own pipe or would broaden its extensive imports of seamless steel pipe for onshore operations to include offshore pipe as well.

With the exception of the Po Hai Gulf and possibly the Yellow Sea, there was no firm evidence as of 1977 that China was moving into production in the other places where offshore operations were known to be under way, notably the area between Tsingtao and Shanghai in the East China Sea, the Tonkin Gulf, and the areas near Hainan [58] and the Paracel Islands [59] in the South China Sea. However, extensive seismic surveys and some test drilling have gone on since at least 1974 in all of these areas, and production is expected to begin in one or more of them in the relatively near future. By 1980 China's combined offshore production of oil and gas in the Po Hai Gulf and other areas is likely to reach at least the 21 million tons (157.5 million barrels) per year anticipated by a group of British experts [60] and could go considerably higher. Jan-Olaf Willums has predicted a slightly higher figure of 28.2 million tons (212 million barrels) by 1980, even if Peking attempts to develop its offshore program without any further inductions of foreign equipment or expertise, followed by a subsequent growth to 74.4 million tons (558 million barrels) by 1985. Depending on the extent of its foreign equipment purchases and arrangements for the transfer of know-how, as Willums points out, Peking's offshore potential would correspondingly increase. Thus, listing eight "technology options" in ascending order, he concludes that China is most likely to choose the third on the scale, which would entail the purchase of three additional offshore rigs abroad by 1978 over and above those already ordered, including one deep-water drillship or semisubmersible. This would result in offshore production of at least 43 million tons (322.5 million barrels) of oil and gas by 1980, he states, climbing to 154 million tons (1.2 billion barrels) by 1985. In contrast, by opting for all-out foreign help, Peking could achieve production levels of 45 million tons (337.5 million barrels) by 1980, followed by a surge to 232 million tons (1.7 billion barrels) by 1985 as initial efforts necessarily focused on exploration are succeeded by steadily intensifying production activity.[61]

The analysis in this chapter would suggest that off-shore production levels of 154 million tons (1.2 billion barrels) by 1985 and 200 million tons (1.5 billion barrels) by 1990 are realistic possibilities for China if one assumes a modest, in-termittent influx of foreign technology comparable to that since 1966 and if Peking gives oil production the budgetary priority envisaged in chapter 2. Indeed, in order to reach its 400-million-ton-plus (3-billion-barrel-plus) target by 1990, China would be compelled to aim at offshore levels of this magnitude. As chapter 2 shows, oil production in northeast China is not likely to go beyond a plateau of 100 million tons (750 million barrels) annually and China would, therefore, have to turn to other areas for some 300 million additional tons (2.3 billion additional barrels) per year. If one assumes that onshore production in the interior areas and in south China rises to 90 million tons (675 million barrels) or more by 1990, offshore production levels of, let us say, 130 million tons (975 million barrels) per year in the Po Hai Gulf and 80 million tons (600 million barrels) in other areas would put Peking over its target. However, it should be emphasized that such high production levels could not be achieved much before 1990, in all likelihood, given the long lead time be-tween the start of intensive exploration and the establishment of a production infrastructure.

China is likely to make its quantum leap in the volume of offshore production in the late 1980s as the groundwork now being laid begins to pay off. In developing the first Japa-nese offshore field at Aga, discussed in chapter 7, Idemitsu Kosan, aided by Amoco, began concentrated seismic surveys in 1968; started test drilling in 1971; made its first discovery in January 1973; put up its first production platform in August 1974; laid its pipeline from April to August 1975; and was ready to go into production four months later.[62] By all indications, China had reached the discovery stage in some parts of the Po Hai Gulf as of 1976 but was still conducting seismic surveys and test drilling operations in most of the gulf and the Yellow Sea. Western and Japanese oilmen also point out that Idemitsu Kosan had technical help from Amoco and that it might take China, on its own, longer than three years to go from the discovery to the production stage. Moreover, these experts add, production projections must take into account the fact that it usually takes two to four years before maximum efficient output rates are reached in offshore production. The Aga field is relevant in studying the Chinese potential because it is located ten miles offshore and involves pipeline and undersea logistic problems comparable to those faced by China in the close-in areas where its first production efforts are concentrated. In contrast to the shallow Po Hai Gulf, however, the Aga site required test drilling at a depth of 210 feet with a semisubmersible rig, posing a more complex technological challenge than China will confront at the exploration stage until it moves beyond the gulf into deeper waters.

In considering the sea boundary disputes emphasized in this book, it should be clearly recognized that China's long-term offshore plans have been designed to embrace contested deep-water areas and that Peking already has a capability for conducting surveys and test drilling operations in many of these areas. According to Peter Hood, leader of the 1974 Canadian mission mentioned earlier, Chinese officials, shopping for navigation and survey equipment, spoke of "a program to explore their continental shelves and deep ocean for petroleum for an offshore distance of about 300 nautical miles." [63] As it happens, 300 miles is the average width of the continental shelf in the East China Sea, which means that the CGG survey boat *Lady Isabel* (now the *Liao Yuan*), with its U.S.-supplied satellite navigation equipment, could conduct seismic surveys in most of the shelf areas claimed by Japan and Taiwan. Given the apparent limitations of its data-processing facilities, China appears unlikely to complete a preliminary assessment of its huge continental shelf before 1979. By that time, however, Peking will have experience in using its new Singapore-made, U.S.-equipped jack-ups, capable of drilling in 275 feet of water. Additional domestically manufactured rigs are also likely to have been completed, and Peking might have a prototype of a semisubmersible.

Although China may not be able to engage in intensive deep-water activity for a number of years, Peking will have the capability in the interim for the selective, symbolic deployment of its rigs in politically desired areas. Most of China's neighbors have unilaterally allocated concessions in areas where Peking has implicit or explicit claims, and sooner or later China seems likely to respond to this affront in its own fashion. Military action would be at variance with the general tenor of post-1972 Chinese policy in Asia, although it cannot be ruled out, and diplomatically, as we shall see in a later chapter, China seems reluctant to give serious status to its offshore disputes by entering into formal negotiations. Peking seems to be playing for time, in short, until its own offshore capabilities are more fully developed. In a few short years China will be able to signal its intentions in specific areas of the continental shelf with survey ships and rigs rather than gunboats. Just as its rivals are now seeking to stake their claims by doing seismic studies and drilling wild-cat wells, so China will be increasingly equipped to do the same as its rig imports and rig-manufacturing program gain momentum.

CHAPTER FIVE

Offshore Oil
and the
Future of Taiwan

The potential impact of offshore oil discoveries on the economic viability of a non-Communist Taiwan or South Korea makes these cases inherently more volatile than the narrower disputes over sea boundaries pending between China and its Communist neighbors, North Korea and North Vietnam, or even between China and Japan. In particular, the destiny of Taiwan is likely to be critically influenced by the economic and political consequences of continuing offshore exploration efforts in continental shelf areas claimed by China and Japan. Would major discoveries provoke serious conflict with Peking, Tokyo, or both? Or is oil the lubricant that could draw Peking and Taipei together? Is Taipei correct in its judgment that Peking views Japan as the common rival of all Chinese and is ready to tolerate Taiwan's offshore claims as a convenient, indirect way of forestalling a Japanese presence on the continental shelf?

Since 1966 Taiwan has proceeded on the assumption that offshore oil and gas may be one of the keys to retaining its independence from the mainland and has successfully enlisted the collaboration of six American oil companies in its exploration efforts. Even if oil is not discovered, Taipei has reasoned, American involvement will help to prolong American support for some form of separate Taiwan, and if there should be major discoveries, this would reduce a crushing oil-import burden that soared to $850 million in 1975. The American companies, for their part, have not only been attracted by the possibility of offshore discoveries but have also helped Taiwan in order to protect their other lucrative investments and markets on the island.

Originally positive in its attitude toward the involvement of American oil companies in Taiwan, the U.S. government has been more ambivalent since the 1970 *Gulfrex* case, discussed earlier, and has actively intervened to prevent drilling in disputed waters in a series of 1975 incidents to be related in this chapter. Taiwan has responded to these U.S.-imposed restraints by embarking on a stepped-up exploration program but was moving cautiously, as of 1977, torn between drilling on its own in politically sensitive concessions, regardless of U.S. wishes, and limiting its offshore operations to concessions close to the island in the hope of minimizing political consequences.

Gambling on Independence

The origins of Taiwan's interest in offshore oil exploration date back to the first appearance of studies by Emery and Niino in 1961 and the subsequent establishment of a United Nations agency dedicated to determining the nature and extent of the offshore petroleum resources in East and Southeast Asia. Known as the Committee for the Coordination of Joint Prospecting for Mineral Resources in Asian Offshore Areas

(CCOP), the new oil agency was largely the creation of C. Y. Li, an enterprising Chinese geologist on the staff of ECAFE [1] with close ties to Taiwan. [2]

The United States and Britain were cool to the CCOP idea from the start, Li recalled, partly as a reflection of a visceral hostility on the part of Western oil companies to the notion of a public intergovernmental agency "meddling" in a sphere where, it was felt, private enterprise had the experience and know-how to do the best job. Li promoted the CCOP concept with missionary zeal, however, not only as a geologist who believed in the petroleum prospects of the area but also as an ambitious bureaucrat who perceived what he regarded as a genuine need for a potentially powerful new organizational entity. Taiwan and South Korea were quick to recognize that the new agency could be used to serve their interests and gave Li their strong backing. Somewhat later, Japan also became one of the early promoters of the new body, which ultimately embraced nine countries. The creation of the CCOP in late 1965 came at a time when the major oil companies were not yet actively interested in the offshore oil potential of East Asia. In some cases, companies were conducting quiet preliminary assessments on their own but were in no hurry to enter into negotiations with the governments concerned. For Taipei and Seoul, especially, the CCOP provided a way of acquiring independent geological information that emboldened them to take the initiative in approaching foreign collaborators and, to some extent, strengthened their bargaining position.

The first foreign oil company to take a serious interest in Taiwan was the American Oil Company (Amoco), a division of Standard Oil of Indiana, which signed a "letter of intent" in mid-1967 envisaging aeromagnetic and seismic studies as a prelude to formal negotiations on a concession. Although the Emery survey discussed in chapter 3 had not yet taken place, there was an atmosphere of growing expectancy in Taipei during this period, as exemplified when the Fourth Session of the CCOP met there amid an enthusiastic official reception. Economic Affairs Minister K. T. Li declared in his inaugural address:

With huge reserves of oil and natural gas now being exploited in the offshore areas of many parts of the world, it is not unreasonable to expect that substantial reserves of petroleum will be found under the seas in the vicinity of the island of Taiwan. Because of the limited amount of mineral resources on this island, we have realized early the importance of developing natural resources beneath our coastal waters.[3]

By the time that Emery did make his CCOP survey in the fall of 1968, Gulf had also begun to conduct preliminary negotiations with Taipei, and the survey vessel *Gulfrex* made a brief appearance in the Taiwan Strait just as Emery's survey ship, the *R. V. Hunt*, was departing from the area.[4] Amoco then followed, in turn, with its promised aeromagnetic survey in early 1969. In an incident presaging the later *Gulfrex* encounter with Chinese gunboats related in chapter 1, the Amoco plane went on its mission with armed Nationalist Chinese soldiers aboard and was shadowed by a Chinese plane throughout its operations. Nevertheless, despite this show of Peking's hostility, Amoco and Gulf became increasingly eager to conclude concession arrangements with Taipei.

It should be kept in mind here that Taiwan entered into its negotiations with foreign oil companies in 1969 and 1970 against the background of continuing Sino-U.S. estrangement and nominal American diplomatic recognition of Nationalist Chinese claims to jurisdiction over the mainland. Taipei had demarcated its projected offshore concession areas, accordingly, not in its capacity as a government with jurisdiction over Taiwan but rather as a regime purporting to speak for China. Invoking Law of the Sea concepts applicable to China as a coastal state, as I shall elaborate in chapter 9, Taiwan had staked its claim to virtually all of the Taiwan Strait and the East China Sea, stopping short of Yellow Sea claims in deference to its anti-Communist Asian neighbor, South Korea, but overlapping substantial areas claimed by

Japan. As initially offered to foreign companies, its concession areas embraced the entirety of the Taiwan Strait up to
the Chinese coast, ranged all the way from the mainland to
the midpoint of the Okinawa Trough, the undersea canyon
separating the edge of the continental shelf from Japan's Ryukyu Islands, and reached northward to a point in the East
China Sea opposite Shanghai on the west and the South
Korean and Japanese coasts to the northwest and northeast
(Figure 5). Like its claims to the mainland proper, these boundaries have never been formally disavowed,[5] but in its actual
concession agreements Taipei has never allocated the full
areas originally staked out. While its concessions do cover the
full distance to the north, they are bounded on the west by
an unacknowledged,[6] hypothetical median line in which
Taipei and Peking are implicitly treated as coequal entities in
international law.

The agreements signed by Amoco and Gulf on 21 September 1970 reflected a mood of political caution that was not
displayed by three smaller companies (Oceanic, Clinton, and
Texfel) in their agreements. Both Amoco and Gulf chose concession areas close to the island in the belief that this would
minimize the danger of collision with China (Figure 5). Significantly, however, both companies assumed that Taiwan
would remain independent of the mainland and were not
afraid of losing their investments as a result of the possible
future accession of the island to Peking.

"The future of Taiwan at that time wasn't as questionable as it is now," Thomas Caffey, former associate general
counsel of Amoco, recalled in 1975. "No one ever thought the
U.S. would turn around and do what we did do in 1972. But,
of course, everybody realized it might take some fighting at
some time to settle the question." Another former Amoco official involved in the negotiations for the Taiwan concession,
Robert Pleasant, also pointed out that

the U.S. position was so strong in Taiwan then that everyone concerned had great confidence. Indeed, given the extent of the U.S.

commitment at that point, it seemed to be a very safe place. There was never any question in anyone's mind that Taiwan was not firmly established as a sovereign power.

Pleasant added that Amoco was only required to make an initial investment of $200,000 and that it was "not even necessary to make long-range political judgments" in order to justify a commitment of this limited character. Taiwan was liberal in its terms, Pleasant noted, because it was "seeking to get a hold on the United States politically" in the form of a U.S. investment stake in the island. Taipei did not ask for the special advance payments, known as signature bonuses, exacted by many governments, and did not emulate the "production-sharing" demands then being made by nearby Indonesia. Moreover, Amoco was seeking to "build a relationship" with Taipei in order to obtain a market for its Iranian crude and to open up opportunities for investment in petrochemical manufacturing facilities. If offshore oil discoveries were made, so much the better; but in any event, Amoco reasoned, the good will produced by offshore cooperation could also bring a payoff in other ways, a calculation that proved to be well founded.[7]

Amoco selected its concession sites after consulting Law of the Sea expert Carl McFarland, a professor at the University of Virginia Law School, who advised against choosing sites claimed by Taiwan on the basis of its "unrealistic or controversial" assumption that it ruled the mainland. In a letter to Caffey on 20 November 1970, McFarland pointed to the Taiwan Strait as the area where Taipei could most plausibly seek to enforce a median line without reference to its mainland claims. The company chose four out of the eleven offshore blocks offered by Taipei on the Taiwan side of the strait, Pleasant recalled, with the basic objective of "keeping as close to Taiwan as possible." McFarland had also advised that the Japanese claim to the Senkaku Islands (Tiao-yü T'ai) northeast of Taiwan might prove viable in international law, which could give Tokyo a territorial position on the continental shelf (as explained in chapter 9) and could thus invalidate Taiwan's concessions to the north of the island irrespective of its boundary conflicts with Peking. For this reason, Amoco steered clear of areas north of Taiwan, including the relatively close-in areas chosen by Gulf.

Like Amoco, Gulf selected its concession site on the basis of an implicit assumption that Taiwan would be able to retain its independence for an indefinite period. In Gulf's case, however, Northcutt Ely, the company's adviser on sea boundary issues, allowed for a wider range of political contingencies than McFarland had done in his recommendations to Amoco. Ely pointed out that a median line between the island and the mainland did not necessarily have to be interpreted as an international line between two sovereign entities but could also be treated as an interprovincial line of demarcation in deference to the "one China" concept upheld by both Taipei and Peking (Figure 11). In an advisory opinion for Gulf presupposing some form of independent status for Taiwan, Ely declared:

If Taiwan and mainland China are to be considered as two nations (which each denies) . . . a median line, in our opinion, is the proper boundary. We refer to it as an "administrative line," which, from the viewpoint of the Republic [Taiwan], is an internal boundary between the province of Taiwan and the provinces of the mainland. Substantially all of the Gulf concession is on the Taiwan side of this line. Whether mainland China would respect such a line is beyond our ability to predict.[8]

Glen W. Ledingham, former vice-president for exploration of Gulf's Asia division, who negotiated the 1970 concession agreement, explained that the company saw the concept of an interprovincial median line as potentially useful in the context of any future political accommodation between Taipei and Peking. Gulf was not concerned with the modalities of Taiwan's relations with the mainland, as such, he said, so long as the island kept the substance of economic autonomy. However, Washington lawyer Donald Allen, who worked for

Ely in 1970, said that Gulf had taken it for granted that Taiwan would remain completely independent. Interviewed in 1974, Don Thomson, then in charge of Gulf's Taiwan operations, felt it was "likely that an accommodation will be reached permitting this to be an independent country"; and Melvin Konts, exploration manager of Gulf in Taiwan, predicted a "Mexican stand-off" or a "Hong Kong-type arrangement."

In choosing the East China Sea concession nearest to the island, Gulf drew a clear distinction between what it saw as the legitimacy of Taiwan's jurisdiction over areas close to its shores and the more dubious validity of Taipei's northernmost concession zones (Figure 5). Ely wrote that with respect to the area of the East China Sea continental shelf north of the thirtieth parallel, "the standing of the Republic of China must be based on its territorial claims to the mainland, not merely to Taiwan." Exploration in the northernmost zones would be a direct provocation to Peking, and these zones were thus inherently undesirable from Gulf's standpoint. Moreover, Ely argued, in view of Taipei's tenuous claims there, a Gulf concession north of the thirtieth parallel would create even more complications in the company's relations with Japan than a concession adjacent to the island. An independent Taiwan, he declared, would have a clear legal case against Japan with respect to areas south of the parallel, and

The present dangers to the Republic (and its concessionaires) in this southern area are primarily political. Accordingly, it may well be to the advantage of the Republic of China and Japan to arrive at an accommodation which will align them together in the protection of their claims to the seabed as against mainland China, perhaps in the form of the joint sponsorship of oil concessions.[9]

By contrast, Ely added, such a Taipei-Tokyo "condominium" would not be possible in the northern zones, where Japan would have greater reason to fear a potential conflict with Peking.

For Gulf it was a matter of some importance to harmonize its interests in Taipei with those in Tokyo, since the company has long had an extensive stake in Japan in the form of crude sales and an exploration partnership with the Tei-koku Oil Company, involving concession areas that overlap some of Taiwan's concession zones. Nor did this seem difficult in the political climate of 1970. Tokyo still had close ties with Taipei then, and Gulf saw no threat to its Japanese interests in Taiwan concession boundaries that embraced the Japanese-claimed Senkaku Islands (Tiao-yü T'ai) and extended to the midpoint of the Okinawa Trough.[10] It appeared both logical and possible that some form of joint concession arrangements would be worked out. As chapter 7 shows, however, Japan's approach to Taiwan shifted abruptly in 1971. Gulf has subsequently attempted to minimize any potential conflict between its interest in its Taiwan concession and its larger stake in Japan, first by formally relinquishing the most sensitive portions of its Taiwan concession near the Senkakus (Tiao-yü T'ai) and later by refraining from drilling operations in those remaining portions of the concession that overlap with other areas claimed by Japan (Figure 6). This has left the company with a relatively limited area within which drilling has been politically feasible, and only four wells had been drilled there by the end of 1976, all of which were dry holes, although said to be geologically promising with respect to surrounding areas. These wells were concentrated in the western corner of the concession zone, less than 115 miles from the Chinese coastal city of Fuchow, which posed less of a risk to Gulf's interests than drilling in areas that might have offended Tokyo. The company has become increasingly cautious in its approach to Taiwan, in any case, amid signs of growing U.S. government interest in completing the normalization of relations with Peking. Still, Gulf's concession agreement is not scheduled to expire until March 1980, and it has a major stake in crude sales ($225 million in 1974) and various investments (e.g., a controlling share in China Gulf Oil, which provides the Chinese Petroleum Corporation with all of its lubricating oil) that could lead to a continuing offshore role as part of an overall effort to keep on good terms with Taipei.

Compared with Gulf, Amoco has had extremely encouraging geological results in Taiwan. In one of its two concession areas, Amoco entered into a partnership with the Continental Oil Company (Conoco), which made two natural gas strikes sixty miles off the southwest coast of the island in October 1974 and November 1975 (Figures 5 and 10).[11] Studies indicated the probable presence of reserves totaling 300 billion cubic feet of natural gas (50 million cubic feet a day for twenty years). For Taiwan, this was sizable enough to make it worthwhile to pipe the gas to shore for industrial uses that would relieve the pressure for oil imports and enhance the economic self-sufficiency of the island. For the two majors, however, much bigger reserves would have been necessary to justify development costs expected to range between $600 million and $1.8 billion.[12] The principal customers for the gas were to be in Taiwan itself,[13] and the government-regulated price for the gas envisaged by the Chinese Petroleum Corporation was not nearly as high as the companies wanted. Despite the value of the discoveries, therefore, and estimated exploration expenditures of $9 million, Amoco and Conoco finally decided against going ahead with development. While retaining much of their concession area, they relinquished the site of the gas field in late 1976 to the Taiwan government, which has subsequently announced plans to proceed with development on its own.

Reflecting on the negotiations leading up to the return of gas field, A. A. Savage, Conoco's manager of acquisitions, said that "political considerations" had played a crucial part in the calculations of both Taipei and the companies. From Taiwan's point of view, he said, "they attach great importance to self-sufficiency, and a governmental enterprise doesn't have to measure profitability in the same way that we do. And there is a lot of gas there." From Conoco's point of view, the future status of Taiwan had begun to look increasingly uncertain in the context of a shifting U.S. China policy.

Washington was unwilling to guarantee the investment as it did many other investments on the island.[14] Continuing American nervousness with respect to drilling in the Taiwan Strait had been shown in the State Department's eleventh-hour intervention in January 1976 to prevent Conoco's use of a U.S.-flag drilling rig. Even if they had started to drill production wells immediately, Conoco and Amoco officials pointed out, and even if a price compromise had been reached with Taipei, it would have taken at least eight to ten years to get into operation and produce enough gas to cover the enormous investment involved. With prospects for normalization between Washington and Peking gradually improving, it no longer seemed safe to assume that the island would retain its economic autonomy for another decade or that the United States would back up Taiwan in any boundary conflict with the mainland.

Fears of a possible dispute with Peking over development of the gas field have a special pertinence in view of its location. Even though the area envisaged for production falls clearly within the Taiwan side of a hypothetical median line with the mainland, it could invite controversy for two reasons. One is that the area falls on the mainland side of a fifteen-mile undersea canyon, and the canyon could conceivably be invoked as a basis for demarcation of the strait. More important, while the gas field itself lies on the Taiwan side of a median line, it is geologically linked with a sedimentary basin that reaches across the line, the Luichow Basin, depicted in a map issued by the Chinese Petroleum Corporation itself as a northeast-southwest basin running from the area where the gas discovery was made to waters north of Hainan (Figure 10).[15] Other geological studies confirm the direction of the Luichow Basin.[16]

The danger of a collision with China resulting from the location of the gas field has been discounted on the argument that Peking is not likely to recognize a median line, in any case, and would only make an issue of offshore oil development by Taiwan if and when it is prepared to press its claim for Taiwan as a whole. This may well prove to be true, but it is also possible, as Amoco and Conoco appear to have concluded, that production platforms and gas pipelines in the middle of the strait could themselves become a focus of factional controversy in Peking and help to bring the Taiwan issue to a boil.

The Clinton Caper

Reflecting its claim to jurisdiction over the entire mainland, as we have seen, Taiwan not only granted close-in concessions but also allocated offshore rights over a 500-mile belt to the north of the Gulf zone that conflicted with Chinese and Japanese claims alike. Given the built-in risks in the northern concessions, they were shunned by the major oil companies, but smaller and more adventurous firms specializing in exploration were attracted by the promising geological prospects of the areas concerned. All three of the most sensitive concessions to the north of the island were granted to relatively small U.S. companies lacking the financial capability to conduct expensive deep-water drilling operations on their own. It is standard practice in oil exploration for such small promoters to locate promising possibilities, spend a few million dollars on surveys, and then try to find a better-heeled partner to put in the multi-million-dollar investment needed for sustained drilling operations. In the Taiwan case one of these companies, Clinton International, made a serious attempt to arrange drilling operations that ballooned into a major diplomatic *cause célèbre* in 1975.

Until 1970 the Clinton Oil Company of Wichita had no international investments and would have seemed an unlikely candidate for one of the more daring offshore exploration gambles then beckoning anywhere in the world. Clinton's oil properties were almost all located in Kansas, Texas, and Oklahoma. Known in the oil business as a tax-haven company or drilling fund, it offered wealthy investors in the 50 to 70 percent bracket a way to offset their income from other sources with the losses resulting from high-risk exploration ventures organized on a "limited partnership" basis.[17] Clinton was one of the more successful companies of this character and was a natural target for a high-powered promoter seeking to draw the firm into international wildcatting ventures.

The promoter in this case came with impeccable credentials. Robert B. Anderson was not only a respected former secretary of the Treasury who had been mentioned by Dwight Eisenhower in his memoirs as a figure of presidential caliber, but also had a successful business record in Texas prior to joining the government. Following his departure from Washington, Anderson had traded on his worldwide contacts, arranging for the sale and purchase of companies in exchange for a fee and often for an equity position in the ventures concerned. Clinton made Anderson chairman of its newly formed overseas arm, Clinton International, in the hope that he would help the new firm get important concessions abroad and would attract U.S. government financial support. As his critics in Clinton bitterly charged when he broke with the company in a flurry of legal battles,[18] however, Anderson's only contribution was the acquisition of what proved to be a highly speculative concession from Taiwan, and he had little to do with the North Sea concession that was to make Clinton International profitable.

In view of the subsequent political complications that have clouded the future of the Clinton concession, it is interesting to recall the strenuous diplomatic efforts made by Anderson to obtain it, including a series of Taiwan visits to plead his case personally with Generalissimo Chiang Kaishek. Aided by an influential member of Chiang's inner circle, Jen-zen Huang, who was paid an $85,300 fee, Clinton won a concession in one of the East China Sea areas that had been identified as most promising in the Emery report.

Huang is a brother of the late General Jen-lin Huang,

who served as deputy commanding general of the Combined Service Forces in the early days of the Chiang regime on Taiwan and later played an influential role as the head of Chiang's personal "peace preservation group," or security force. While "J. L." Huang remained a power in Taipei, "J. Z." Huang had come to Washington as a troubleshooter for the Generalissimo. "J. Z." once boasted that he played poker with President Truman and Speaker Sam Rayburn, and during the Eisenhower years he had become acquainted with Anderson. In its 1950 exposé of the "China lobby," *The Reporter* magazine described Huang as the ever-present right-hand man of Louis Kung, "courier and paymaster of the China lobby." [19] It was Kung, in turn, who was identified as the key figure in the China lobby's active financial support of Richard Nixon in his 1950 Senate race, as well as in many other key Senate contests in that year.

The fact that J. Z. Huang had received $85,300 as a middleman in arranging the Taiwan concession was buried deep in a registration statement filed with the Securities and Exchange Commission (SEC) in March 1971, when Clinton International decided to put an $11-million stock offering on the market. [20] The offering was canceled after it proved impossible to find an underwriter who would back it, but the registration statement remained on file and provides detailed information on the terms of the Clinton concessions in Taiwan and elsewhere and the personnel involved. [21] Huang had not only received his initial fee, the statement disclosed, but was also to receive an interest of 1 percent of any production, which the company could buy back for $1.5 million. [22] The statement further revealed that Anderson had invested only $20,000 in return for 20 percent of the stock in Clinton International.

In accordance with the SEC requirement that companies spell out possible risks as well as possible gains to prospective investors, the statement noted:

The Taiwan property is located on the continental shelf of the People's Republic of China [mainland China], and is closer in distance to mainland China than to Taiwan. By virtue of this fact and the fact that mainland China presently claims territorial rights to Taiwan and the islands under its jurisdiction, the company believes that mainland China would consider the Taiwan property to be within its jurisdiction. If mainland China attempted to exclude the company from engaging in exploration activities on the Taiwan property, the company's rights under the Taiwan Agreement would be adversely affected to a material degree.

The statement added that the company was "not aware" of any Japanese claim overlapping with the Taiwan concession. This was literally accurate in that Japan had avoided formal continental shelf claims out of deference to China and had not taken Taiwan's claims seriously until Taipei awarded its first concessions in late 1970. By early 1971, however, Japan had begun to intensify its public opposition to Taiwan's shelf claims, especially in the Senkaku (Tiao-yü T'ai) area, and had also begun to stake out "shadow" concessions on the shelf. Japanese companies were permitted to register their applications for specific areas with the Ministry of International Trade and Industry, but the companies concerned were not given formal authorization to explore for petroleum pending a resolution of the territorial issue.

Anderson responded to the Japanese moves by embarking on a sustained campaign to translate his contacts in Tokyo into a Japanese concession for Clinton coinciding with areas where Taipei had granted its concession. As Treasury secretary he had dealt frequently with Prime Minister Sato, and he unsuccessfully sought help from the premier in a series of meetings on 29 October and 1 November 1971. Among other things, Gulf's Japanese collaborator, Teikoku Oil, had already applied for a Japanese concession covering the area concerned. Anderson returned to Tokyo for negotiations with Teikoku in February and April of 1972, but his offer to trade a half interest in the Clinton concession for a half interest in the Teikoku concession was rejected, reflecting the overall shift in Japanese political attitudes that had taken place as a result of the Kissinger visit to Peking in mid-1971 and the Nixon visit to China in early 1972. Japan was already preparing for its own policy shift from a pro-Taiwan to a pro-Peking orientation. Anderson continued to shop for a Japanese partner throughout 1972, nonetheless, meeting in New York in October with a representative of the Kyushu Oil Development Company, a now defunct company closely linked with Nippon Steel. As late as 1973, Anderson was still trying to borrow money in Japan for drilling in the Clinton concession area, promising repayment in the crude oil to be produced there.

Significantly, while Anderson was making his bid to the Japanese leaders, the managing director of Clinton International, Ted C. Findeiss, was attempting to enlist behind-the-scenes U.S. government backing for collaboration between Tokyo and Taipei in the East China Sea. Findeiss, a bluff, hearty former Republican state senator in Oklahoma, made a direct overture to Ambassador Walter P. McConaughy in Taipei on 15 October 1971. As summarized in a memorandum detailing his presentation to McConaughy, Findeiss told the ambassador that the initial seismic studies in the Clinton concession had been "extremely encouraging, pointing to the possibility of immense oil deposits of Middle East size." The implications of a discovery could be "profound," Findeiss argued, since "Japan would get oil where she needs it, the Republic of China would get credibility, and the United States would be less dependent on the Middle East." Clinton would go ahead on its own whether or not a deal could be worked out between Japan and Taiwan, he said, but the United States should try to bring the two together because "it's so logical for these governments to get together that there should be no question in view of the tremendous benefits to the Republic of China, Japan, the U.S., Western Europe and the whole world, especially the Free World. If two allies of the U.S. could work together, this would obviously strengthen their case against the mainland." In addition to promoting negotiations between Tokyo and Taipei, Findeiss suggested, Washington should permit Clinton to use satellite navigation in operating its seismic survey boats while the negotiations are in progress, "since the bargaining process could stretch out over a long period." [23]

Speaking in a barrage of superlatives during a London interview in late 1974, Findeiss spread out his seismic studies and maps on the table, pointing first to one particular undersea geological configuration in the northeast corner of the concession some thirty to forty square miles in size and then to eight others scattered nearby (Figure 4). "That's the biggest damned structure you ever laid eyes on," he said, and in geophysical terms, the odds are "ten to one" that the concession as a whole would yield at least 4 billion tons (30 billion barrels) of oil, 1.7 billion (12.8 billion) in the one best structure alone. Even if the yield proved to be only 1.3 billion tons (10 billion barrels), he calculated, this could bring $100 billion if oil sold at $10 a barrel or more. Assuming a maximum investment of $5 million in order to find oil, this would be a 20,000-to-1 shot, and "there ain't no horse race with that kind of odds." Once the oil had been discovered, he went on, "I'd ask Mao and Chiang to come to the same table," and if Peking proved unwilling to agree to a division of the spoils with Taipei,

then we would get out the F-104s and fight. This is go, no-go poker. Of course, there will be a lot of hair-pulling if we find anything, but do you want to see it all go to the Chinese? Unless of course you up and give all of Taiwan to the Chinese. Remember, if the Chinese own that, they own everything on Taiwan, those airports, those hotels, everything.

Findeiss predicted that "Red China would raise a lot of hell in the U.N. if we start drilling out there," but would not take military action because a military response would be logistically difficult for Peking in the middle of the East China Sea, far from shore, where the Clinton concession is located. "I'd be more scared in the Taiwan Strait," he explained, since "they can lob surface-to-surface missiles against our rigs for target practice there."

Findeiss had invested substantial money of his own in the Taiwan concession in the confident expectation that the island would remain independent and that "both Peking and Taipei would have a place in the sun. If you assume that Taiwan is a separate country, then that means a median line and our concession falls on the Taiwan side of the line." Even if Peking did take over the island, he suggested, "perhaps we could stay on as a contractor—paid in kind, not in cash."

Armed with the encouraging geological data resulting from the company's $1.1-million investment in seismic studies, Findeiss had gone from one oil company to another in the United States and Japan in search of a partner able to invest the estimated $5 million needed for a meaningful drilling effort. His seismic studies had evoked keen excitement, but this was offset by uncertainty as to how China would react and by the growing pro-Peking shift in U.S. policy toward China. The cautious reception to Clinton's overtures was accentuated by an intramural legal imbroglio involving the parent company that had resulted in considerable damage to Clinton International's reputation as well as the withdrawal of Robert B. Anderson from all but a token interest in the company.[24] In early 1974, however, Findeiss finally found a highly desirable collaborator among Clinton's partners in the North Sea, Superior Oil, one of the larger and more solid independents operating abroad.[25] Superior's founder, William Keck, was a legend in the oil industry, and the company still retained some of the gung-ho spirit of its earlier, wildcatting days under H. B. Keck, son of the founder. The Houston-based company was attracted by the promising geology of the Taiwan concession and decided that it was worth taking some chances to get an inside track on the thirty-square-mile structure in the northeast corner of the concession. To the extent that political factors influenced Superior's attitude, they were a plus, since the fact that the oil would help Taiwan to assure its independence appealed to some Superior executives. One leading oil geologist with a close knowledge of the Superior management said that they were "unsophisticated internationally and basically Goldwater-ites." Others in Houston explained the Taiwan decision as true to form for a "gutsy" company that was "willing to take a chance." J. B. Harrison, president of Zapata Exploration, recalled that he had turned Clinton down before the concession was offered to Superior. "Frankly, I didn't have the guts," he said. "Or you might say I saw the future differently." Superior had joined Clinton, he had heard, because "they believed Taiwan would be in the fold."

When I talked to Findeiss in December 1974, Superior was actively looking for a rig but had begun to recognize, for the first time, that it had a tiger by the tail. Superior's lawyers were investigating the international law factors involved in the Taiwan concession, the attitude of insurance companies toward drilling there, and the way that Peking and Washington might react to drilling. "All systems are go, but we might blow up before we get off the runway," Findeiss commented. By early February 1975, when I visited Houston, Superior had actually lined up a rig and was scheduled to drill in April and May but had agreed to discuss the matter with the U.S. government. The State Department had indicated its interest in discussing the political factors involved, and my Houston visit came precisely one week before scheduled meetings on February 11 and 12 at the State Department and the White House. Joseph Reid, president of Superior's international division, explained that the company was seeking to balance political and geological considerations. "We recognize that this carries political question marks," said Reid, "but a political dry hole is not a lot worse than a geological dry hole." Pointing on a map to the site where Superior planned to drill (Figure 5), he added that the company had been surprised to receive a letter of protest from the Teikoku Oil Company and wanted to know if it was true that Japan also claimed jurisdiction over the area concerned in addition to Peking.

In Washington, Reid, Superior International's assistant general counsel Richard Dye, and a consultant, Stanley Lubman, conferred at the State Department with Oscar

Armstrong, director of the Office of the People's Republic of China Affairs, and other government China specialists. One of those present at the February 11 meeting said that Reid and Dye were aware that U.S. policy had long been to discourage drilling in disputed areas by denying licenses for the equipment needed to make use of government satellites for navigation purposes and by forswearing any responsibility for the defense of U.S.-owned rigs or survey ships. What they were not prepared for was the grim report on the current Peking attitude that they heard from the State Department officials. They were startled to learn that Peking laid serious claim to their concession area, had a case with some standing in international law, and was preparing for an offshore program of its own. They were cautioned that if they persisted, Secretary Kissinger would take the matter up personally with Keck. Reid asked about the possibility of "parallel contracts" with Taipei and Peking. "They were extremely naive," a State Department participant recalled. "They had the idea that 'We'll go to the Chinese Communists and work something out. We'll give them a share.' " After conferring at the State Department, Reid and Dye met with Richard Solomon, then the ranking China specialist on the staff of the National Security Council, and encountered an even firmer attitude.

Back in Houston, Superior was hesitant to renege on such a geologically attractive deal, despite the doubts aroused by the static from Washington and Tokyo. The company was finally deterred from its drilling by a last-minute complication, indirectly linked with events in Vietnam, that reinforced its political anxieties. Since sophisticated, semisubmersible rigs cost some $50,000 per day to operate, and since it was so uncertain how China would react, Superior had been prepared to drill only one well and had arranged to subcontract the deep-water *Margie* from Shell. Shell had planned to use the rig for three wells off Vietnam, subleasing the rig for a year from Atlantic-Richfield, but made an unannounced decision in early February to cut back on its Saigon commitments and canceled its contract for the *Margie*. At that point, in order to get the rig Superior would have had to take over Shell's costly, long-term contract.[26] The deadline for an answer to Atlantic-Richfield was 3 March, less than a month before the scheduled date for drilling in the Taiwan concession, and the deadline was extended for two weeks while Superior agonized over its decision, shopping around frantically for possible partners who might use the *Margie* for the remainder of the year after the completion of drilling in the East China Sea. It was nip and tuck, according to Paul Ravesies, vice-president for exploration of Atlantic-Richfield, and there is no telling what might have happened if Superior had managed to find a partner. In the end, Superior backed out of the deal amid a cloud of legal recriminations between Superior, Clinton, and bitter Taiwan government oil officials,[27] who blamed it all on Kissinger and talked of drilling on their own if necessary.

Looking back on the episode, John Wagner, chief counsel of Superior International, reflected that "it's good to be imaginative, but there are forces in this world that can take the wind out of your sails." Richard Dye posed the question: "Just how far do you fight the battle? You look at one prospect, it has one thing against it, you add up the good things and the bad things. And then you estimate the costs involved in the light of your worldwide options." One factor that had tipped the scales, he said, was that the cost of drilling in the Taiwan concession had at least doubled in the year since Superior first signed its agreement with Clinton. From the beginning, the company had known that the East China Sea well would be unusually expensive in view of the water depth (310 feet), the wave action, and the difficulty of supplying a midocean rig beyond helicopter range. But costs had rocketed beyond all expectations following the oil-price crisis and the resulting repercussions in the world rig market.

Stunned by Superior's volte-face and the strong stand taken by the State Department, Findeiss and J. Z. Huang shifted their tactics in mid-1975 and attempted to trade their concession for one in another, less controversial area closer to the island. In the absence of another company ready to take it over, however, Taipei balked, and matters were still in a state of suspended animation in mid-1977. What would happen to the concession? "We'll hold on to it as long as they let us," replied Richard Volk, president of Clinton's parent company, Energy Reserves Group, "and who knows, in five years, it may be possible to drill." Findeiss mused:

It might just float away into nothingness, for after all, circumstances alter cases and wise men alter their plans. We are oil finders, and you can't get oil from a law suit, even though Superior owes it to us to drill. But don't forget that the people in Taipei are madder than hell about this. They may very well take it back from us and drill there on their own if they can get somebody to do it on a contract. And someone might just want to test this thing and see whether Peking won't come to the table.

"Behold the Turtle!"

Like Clinton, the other two companies with holdings north of the Gulf zone, Oceanic Exploration and Texfel Petroleum, are both small operators by oil industry standards and were given concessions only because Taipei was unable to get any of the majors or the larger independents interested in these areas. Like Clinton, both were willing to gamble that even if Taiwan could not enforce its own title claims and proceed with actual oil production there, neither could anyone else without taking into account the exploration that had been conducted under Taiwan's auspices. At some stage, they reasoned, they would be able to drill for oil or unload their concessions at a profit to Peking, Tokyo, Taipei, or some combination of the three. Likening companies of this character to scalpers at a football game, Charles Di Bona, executive director of the American Petroleum Institute, observed that "when the game is played, they have the ticket."

At first glance, Oceanic looks like just the sort of gung-ho company that might actually be prepared to test Peking's response to drilling in the northern part of the East China Sea. The company has consciously sought out politically uncertain concessions throughout the world in the belief that this is a necessary avenue for competing with the established majors. An aggressive-looking turtle dominated the cover of Oceanic's 1973 annual report, which bore a motto proclaiming: "Behold the turtle! He makes progress only when he sticks his neck out." Oceanic's founder, petroleum engineer Jack J. Grynberg, is an intense, Polish-born Israeli refugee who started the firm in 1969 after successfully developing oil properties in the United States. Grynberg makes no bones about his anti-Arab, anti-Communist political views, declaring: "We are dedicated, fiercely dedicated I might say, to finding new sources of energy so that the free countries can have their own energy sources. We are dedicated to maintaining the balance of 'oil power' in the world." [28]

Partly as a reflection of this underlying political thrust, Oceanic became one of the very first American companies to take an active interest in obtaining a concession in the East China Sea, recognizing that "if the free countries don't develop that oil, the mainland will." As chapter 10 reveals, there were powerful figures in Washington who tried until the eleventh hour to make the return of Okinawa to Japan contingent on Tokyo's agreement to grant offshore oil concessions to U.S. companies. Hoping to benefit from any such arrangements, Oceanic, then only a few months old, applied to the U.S. military administration in the Ryukyu Islands for a concession in June 1969. Later, when Japan refused to entertain the application and Taiwan put concessions up for bidding in 1970, Oceanic was one of the first applicants.

Even though it was a small company, Oceanic's stature was enhanced, in Taipei's eyes, by the fact that it had acquired a substantial partner, Fluor Drilling Services, a subsidiary of the vast West Coast Fluor construction empire. Fluor was already well established in Taiwan, which may account for the fact that Oceanic was given preference over Clinton when concessions were awarded. Oceanic did not get its first choice, the concession closest to the island; that one was awarded to Gulf, as we have seen, but Oceanic did receive the next closest, relegating Clinton to its location 350 miles away from the island. An interesting footnote in this connection is that Oceanic, like Clinton, found a powerful political ally in Chiang Kai-shek's inner circle, James I. C. Chiang, a nephew of the Generalissimo. All concerned deny that "Jimmy" Chiang had a hand in the concession award, however, and it was not until some months later that he became a senior vice-president of Oceanic Exploration Company (Taiwan), with stockholdings of undisclosed size in both this venture and the parent company. Educated in Texas, where he has lived off and on for twenty years, Chiang came back to Taiwan in 1960 with funds provided by H. L. Hunt ("a strong anti-Communist with whom I had a lot in common") and has since built up a multifarious network of business ventures in oil and other fields.[29]

Grynberg and Oceanic's president, Wesley Farmer, explained that they attempted to obtain the concession ultimately granted to Gulf because they, too, recognized that Taipei would be unable to sustain territorial claims in the East China Sea based on its purported jurisdiction over the mainland. In a required prospectus filed with the Securities and Exchange Commission in October 1971 Oceanic acknowledged:

Even if the territorial independence of the Republic of China is assumed, the People's Republic of China has substantial grounds to claim jurisdiction over approximately 45 percent of the company's concession area, based upon generally accepted principles of international law which divide the jurisdiction over offshore areas between two countries on a median-line basis.[30]

In a later supplement to the prospectus,[31] Oceanic revised this estimate, contending that only 20 percent of its concession would go to China under a median-line arrangement. Nevertheless, in a 1973 research report on Oceanic, Bache and

Company, the Wall Street brokers, noted that while the Taiwan concession was "an exceptional area geologically, it may not be that good politically." The report estimated the value of the Taiwan concession at $5 million.[32]

Under its concession agreement, Oceanic was required to drill a well by March 1974, climaxing the first of two four-year exploration phases. By late 1973 the company had obtained and interpreted seismic data covering some 6,000 miles of survey "lines" and had promised its stockholders that drilling would begin "by early 1975, subject to rig availability." In a reference to reports that Superior planned to drill in the Clinton concession to the north and that Gulf would renew drilling in the zone to the south, the report added that "drilling activity in the East China Sea by all license-holders is expected to increase during 1974." [33] The Gulf, Oceanic, and Clinton blocks were all located in the same geological basin, and the Gulf and Oceanic areas in particular were closely linked geologically. Oceanic wanted to let Gulf drill first, since a strike would have made it much easier to get financial backing. As we have seen, however, Gulf not only drilled dry holes in 1974 and 1975 but restricted its drilling to the portion of its concession area that did not involve an overlap with Japan. This was not the portion immediately adjacent to the Oceanic area, most of which was claimed by Tokyo.[34]

When Oceanic failed to drill in early 1974, Taipei extended the deadline to 21 March 1975, and a 1974 Oceanic report assured both stockholders and impatient Taiwan officials that they were actively seeking drilling equipment suitable for the severe weather conditions encountered in the East China Sea and were meanwhile beginning to accumulate pipe and other supplies that would be required for the first well. [35] The plea that deep-water drilling equipment was unavailable was to become a standard excuse, even though it became increasingly apparent that the real reasons for Oceanic's failure to drill were the stiffening U.S. government stand against drilling in the northern concessions, as reflected in Superior's withdrawal, and the company's fluctuating financial fortunes. Grynberg had become more and more overextended in his multiplying ventures, barely keeping a step ahead of his creditors during the 1973–75 period, and was finally forced to sell his shares in the company.[36] Moreover, Oceanic had become estranged from its partner in the Taiwan venture, Fluor Drilling, as a result of legal disputes resulting frmm partnerships elsewhere, and Fluor sold its interest in the concession back to Oceanic in late 1976.

Faced with Fluor's withdrawal and continuing State Department opposition to drilling in the northern concessions, Oceanic all but gave up serious efforts to get financial backing from U.S. companies in late 1976 and called on Taipei to assume more of the risks that would have to be taken to develop East China Sea petroleum. One proposal made by Oceanic was for a revised concession agreement under which the Taiwan government would put up a major share of the money needed for drilling, partly through selling stock to local businessmen on the island and to overseas Chinese. Ross McClintock, president of Fluor Drilling, took another tack, pointing out that his decision to withdraw was influenced partly by the realization that the high cost of production so far from shore could only be justified by the discovery of a very large field. As it happened, he said, the most promising prospects in his concession were in a twenty-one-square-mile structure that overlapped with the Clinton zone. Taipei's 1970 concession boundaries were "artificial in the first place," and a new, integrated approach to drilling in the East China Sea would have to be organized by Taipei—or Peking. "I don't expect to see that happen in the next few years," McClintock declared, "but someone will do it sooner or later." As a governmental entity, he added pointedly, the Chinese Petroleum Corporation (CPC) could easily be converted into an arm of the Peking regime in the event of a takeover by the mainland, and the companies holding concessions from the CPC at that time "might wake up one day and find that they are working for the mainland."

Small Stakes,
Big Dreams

In one degree or another, the idea that a Taiwan concession might be translated into some sort of arrangement with Peking has been privately entertained by all of the companies involved in Taiwan, but only one company, Texfel,[37] has made efforts to bring about such an eventuality. As the farthest from Taiwan and the closest to Japan and South Korea (Figure 5), the Texfel concession is even more tenuously situated than the Oceanic or Clinton areas. A Texfel official recalled bitterly that he almost got the Clinton area in 1970 until "Anderson put out money to Chiang and pushed us aside." The only area left then was the one to the north of the Clinton concession, and Texfel had just accepted it in August 1971, when Henry Kissinger made his surprise first visit to Peking.[38] According to Irving Schwade, vice-president for exploration, the company recognized that the new U.S. posture toward Peking would make it difficult to mobilize capital for drilling, but felt that preliminary exploration efforts in an area with such high geological promise would pay off in some fashion and invested $900,000 in seismic studies during 1971 and 1972.

It should be noted here in passing that Texfel's owner, David B. "Tex" Feldman, is a flamboyant jet-set promoter, as unconventional in the context of the international oil business as his Clinton and Oceanic counterparts. Like Anderson, J. Z. Huang, Findeiss, and Grynberg, Feldman hoped to parlay relatively small stakes into multi-billion-dollar profits through the right combination of geology and diplomacy. A Dallas junk-dealer who started out by acquiring oil properties in Texas and Oklahoma, Feldman branched out into the international field in 1958, first in Turkey and Libya and later in Latin America. He once staged a $125,000 New Year's Eve party at Romanoff's in Beverly Hills in which he re-created the decor and sets of *My Fair Lady* for 283 film-star and socialite guests. *Life* magazine termed the affair "the nation's most lavish year-end celebration . . . in a setting of Edwardian elegance that would do credit to a period epic." [39]

Unbeknown to Taipei, Schwade wrote to Deputy Undersecretary of State John Irwin on 8 February 1973 proposing that the United States "take an active part—almost waging a diplomatic offensive—in promoting reconciliation between Peking and Taipei. A united China would be good for the world, provided there is a blending of the better aspects of each." Schwade proposed that the United States seek to obtain Peking's recognition of prior agreements made between Taiwan and offshore oil operators, along with land-based businesses, and that the United States encourage and assist in East China Sea boundary settlements between a "united China," Japan, and South Korea.

Specifically, Schwade urged an East China Sea treaty, to which the United States would be a party, declaring that "proper persuasion" should see Taipei softening its stand, and a "fat carrot"—the potential gains in foreign exchange accruing to a united China from offshore oil—"may see Peking do the same." Pending such an agreement, he said, the U.S. government should provide insurance up to $200 million for companies prospecting in the East China Sea. This would be in the U.S. interest, he contended, because there were 1.3 to 2.7 billion tons (10 to 20 billion barrels) of recoverable crude oil there, and developing these reserves would "reduce the OPEC threat by providing a home-owned 'substitute.'" Moreover, it was only fair to consider the plight in which companies with East China Sea concessions had been placed by the shift in U.S. China policy. "Prior to the opening of exchanges with Peking," he stated, the East China Sea had "appeared to offer reasonable expectations that political risks were relatively 'containable,' and in view of the extraordinary exploratory promise, they were certainly worth taking."

In May 1973 Schwade called on Irwin's successor, Willis Armstrong, but received no more encouragement from Armstrong than he had received earlier. American policy, he was told, was simply to discourage drilling in disputed areas.

Four years later, Texfel was still looking for a partner and still pleading with the State Department to back up drilling in the East China Sea. Like Oceanic, Texfel called on Taipei to help share the increased risks that had resulted from the new uncertainty of U.S. China policy. In particular, Texfel urged that the Taiwan government step in where private insurance companies feared to tread, providing insurance against the possible expropriation of rigs by Peking. Taipei, for its part, had a legal basis for abrogating the Texfel concession agreement but was reluctant to do so lest this be construed as a surrender of its title claims to the concession area.

The Do-it-yourself Option

The failure of Clinton, Oceanic, and Texfel to fulfill their contractual drilling commitments left Taiwan with three options in 1976. One was to abandon exploration in the northern concessions in deference to U.S. pressure, letting the concession agreements remain nominally in force but not seeking to enforce their exploration timetable. A second was to seek to renegotiate the agreements, offering to share more of the costs of drilling. Still another was to abrogate the concessions and sponsor and finance drilling operations on its own, either by hiring foreign contractors or by building its own rigs.

As a result of the Clinton affair, the third option was treated seriously for the first time beginning in late 1975. For one thing, American companies were more and more nervous about getting involved in Taiwan, and Taipei saw little hope of finding others to take the place of Clinton, Oceanic, and Texfel or even to undertake serious exploration closer to the island. Atlantic-Richfield, for example, had been seriously interested in a concession close to the island that had not been put up for bidding in 1970 but backed out after Washington intervened with Superior. Paul Ravesies, vice-president for exploration, said that while the geological factors in the area concerned are good, "they are not good enough to outweigh the political risks. The simple fact is that Red China has the best claim to these resources and there are a lot of other places to spend your money. Taiwan is not a good place to fly your flag." Comoro Exploration, a small promotional company, had painstakingly put together a consortium of U.S., Canadian, and West German companies,[40] and had been promised a concession close to the island until the Clinton affair occurred. "The impact was immediate," declared Comoro's vice-president, Robert E. King. "After that, as fast as you sold an interest to one company, some other company already in it would suddenly back out." Moreover, in addition to the uncertainties involved, Taipei began to view its links with U.S. companies as an inconvenient entanglement that made it more vulnerable to pressure from Washington than it would be on its own. In September 1975 the CPC announced that all offshore areas not yet allocated would be explored and developed by the Chinese Petroleum Corporation. Soon thereafter, the CPC unveiled a $1.8-billion, seven-year program for offshore oil and gas prospecting that was to begin with an $82.5-million budgetary allotment in the fiscal year beginning in July 1976 and would then grow progressively more ambitious. Twenty-eight wells were projected in the first three years, sixty in the next two, and seventy to eighty in 1981 and 1982,[41] in contrast to the sixteen wells drilled prior to 1976, nine of them by foreign companies.

In one of the first expressions of its new, independent attitude the CPC tried to find an American rig company that would drill in an unusually promising site within thirty-five miles of the Senkaku Islands (Tiao-yü T'ai) in late 1975. Approaches were made to Fluor Drilling for the *Wodeco IV*, then drilling off the west coast of the island under contract with the CPC, and the Offshore Company, which had a semisubmersible available in Japan. A deal had almost been closed with one of these, U.S. government sources said, until the State Department got word of it and strongly intervened. The issue remained in abeyance until mid-December, when Taipei finally agreed to drop the idea. Meanwhile, Taipei was also seeking unsuccessfully to lease a semisubmersible rig in Australia or Japan that could drill in the Oceanic area under a joint financing arrangement with Oceanic and the CPC. Negotiations were far advanced for leasing the *Hakuryu II* until the rig's owner, the Japan Drilling Company, discovered in late November where it was that Taipei planned to drill.

How far—and how fast—can Taipei actually proceed in offshore activity on its own if it continues to encounter difficulty in hiring foreign rigs for politically sensitive drilling operations? Given its primary reliance on American oil companies in exploration activity, the CPC had not made as much effort as Peking to develop an indigenous offshore program prior to the Clinton affair and would need significant elements of foreign technical help to make and use deep-water rigs in the foreseeable future. Nevertheless, Taipei has substantial capital available for offshore development and has acquired much more offshore expertise than is generally recognized. Ross McClintock, president of Fluor Drilling, said that the CPC already possesses more know-how in the offshore field than the better-known Indonesian state enterprise, Pertamina, and is rapidly upgrading its capabilities. The CPC's level of sophistication can best be understood in the context of its history as the successor to the Chinese Nationalist oil development enterprise on the mainland, also known as the Chinese Petroleum Corporation. When the CPC set up shop on Taiwan in 1949, it had an already established nucleus of well-trained officials and technicians. It then received a quick infusion of Western technology from its new American partner, Standard Oil of California (Socal). By 1975 the CPC had bought out Socal and, as the island's petroleum monopoly, reaped annual profits of more than $100 million. To process its crude imports from the Middle East it operated refineries with a capacity of 8 million tons (60 million barrels) annually, a figure that was slated to reach 12.8 million tons (96 million barrels) by 1977. It had its own tanker fleet, built on the island by the Taiwan Shipbuilding Company. The CPC had ten onshore rigs as of 1974, had drilled wells up to 20,000 feet in depth, and its drilling teams were exploring in the Philippines and Jordan in partnership with government enterprises there. Although initial onshore drilling results had proved discouraging and onshore production was only 200,000 tons (1.5 million barrels) of crude in 1975, a British consulting firm has predicted that Taiwan's onshore and offshore production will reach an overall level of 12.4 million tons (93 million barrels) per year by 1985.[42]

In addition to its onshore drilling experience, Taiwan has gradually been absorbing the know-how needed for offshore drilling operations as well as for manufacturing offshore rigs and production platforms. In 1973 the CPC hired a barge-type rig from Fluor Drilling, the *Wodeco IV*, under a long-term contract providing specifically for on-the-job training for a shadow crew of its own technicians working side by side with the regular crew. By 1976 it had added another jack-up, obtained from a Swedish firm, SCAN Drilling, and had drilled seven wells,[43] mostly in the Taiwan Strait. Fluor officials said that thirty-eight of the fifty-five regular crew members on its rig were CPC employees. One of Fluor Drilling's deep-water rigs, the *Wodeco VIII*, was built by the Taiwan Shipbuilding Company. A self-propelled drillship, the $20-million vessel was converted from an Esso tanker and can drill in 600 feet of water. Although the conversion of the *Wodeco VIII* was supervised by Fluor Drilling, Ross McClintock, Fluor's president, declared that Taiwan Shipbuilding could duplicate it easily if certain components could be imported. In addition to Taiwan Shipbuilding's facilities at Keelung, with a capacity for building 300,000-ton vessels, the China Shipbuilding Corporation opened a $200-million yard at Kaohsiung in late 1976 able to build 455,000-ton vessels. By 1978, a leading U.S. oil trade journal speculated, Taiwan will be "one of the world's leading builders" of ships and production platforms and "possibly" offshore rigs as well.[44]

Oil and Autonomy

Taiwan's attempt to develop its own offshore expertise is part of a larger effort to maximize its self-sufficient economic viability and thus to prevent its incorporation by the mainland or, at the very least, to improve the chances that the island would be able to retain economic autonomy in the event of a future political and military accommodation with Peking. I have had intensive conversations regarding the role of oil development in Taiwan's future, not only with six of the principal concerned officials in the Ministry of Economic Affairs and the Chinese Petroleum Corporation, but also with American officials and well-informed American oilmen long resident in Taiwan. Significantly, many Taipei officials, echoed by the U.S. oil companies in Taiwan, appear to regard oil as a potential lubricant in some form of tacit compromise with Peking that would permit the island to retain its autonomy indefinitely. By the same token, the possibility that petroleum exploration might harden Peking's posture toward the island is generally discounted.

"They have in the back of their minds that they might turn out to be the exploration arm of the mainland," observed Wesley Farmer, president of Oceanic. As many in Taipei see it, Peking will be too preoccupied with its close-in offshore exploration to move into deep-water drilling activities in the foreseeable future. Jerome S. Hu, president of the CPC, explained:

The farther you go from your logistical home base, the more offshore exploration costs, which means that the mainland is likely to work in close-in areas for some time. Once they actually start working offshore on a large scale, they will have more and more difficulties, just as we have had, and they will realize that they cannot go too rapidly. It is very, very difficult to do it alone. In fact, nobody can really do this entirely alone.

In this view, economic logic dictates a division of labor that would permit Taipei to proceed with deep-water exploration while Peking concentrates on its close-in areas. If Peking nevertheless contemplates deep-water activity, its underlying motive is political rather than economic: the desire to preempt Japanese exploration of continental shelf areas claimed by China. But here, too, Hu saw hope for avoiding conflict with the mainland. "Neither of us wants Japan to develop these areas," he declared. "It's basically an issue between us and Communist China, and over time, we hope to be able to come to terms."

K. S. Chang, vice-minister for economic affairs, said that Taipei's tactics and timing with respect to drilling in the northern concessions would be determined by its assessment of the mood on the mainland and denied that the possibility of a Japanese response concerned Taipei: "We are not hesitating because of the Japanese. The only conflict here is between Taiwan and mainland China. But we think Peking would much prefer to have us develop this area, as Chinese, rather than let the Japanese do it, even though we have our ideological disagreements."

Both Hu and Chang minimized the possibility that Peking would choose to make a display of military force against Taiwan over oil exploration as such and argued that any military showdown with the mainland would more likely occur as part of a larger effort to force a resolution of the entire Taiwan issue. For the record, Hu declared that "we are able to defend ourselves. If they blast one of our platforms, we're going to blast them." But he added that "everything depends on not fighting. If we can't wipe them out, we can nonetheless try to live with them." Chang claimed that "if they want military action, they would take it against a military target, probably Quemoy, not against a rig."

While reluctant to admit it, Taipei is keenly aware that its naval defenses would prove no match for Peking in the event that Peking did force a military confrontation over an oil rig or a survey vessel. Chinese Nationalist patrol boats have five-inch guns with a range of fifteen miles, while Peking has Komar-class patrol boats armed with Styx missiles with a twenty-mile, over-the-horizon range, as well as destroyer escorts armed with Riga missiles.

Taipei officials pointedly distinguished between exploration and production and suggested that Peking may see some advantages in letting Taiwan and its foreign collaborators make the costly investments necessary to determine whether—and where—there are extensive petroleum deposits on the continental shelf. If there is a danger of a clash with the mainland, they said, this would arise when and if a major oil or gas discovery is actually made. Asked what they might do at that juncture to work out a compromise with Peking, most Taiwan leaders were vague and cautious, but I did not elicit very strong objections when I pointed out that some American diplomats and bankers on the island talked of possible tax agreements that would give Peking a share of the revenues accruing from any oil or gas produced by Taiwan.

As chapter 11 seeks to show, Peking's reaction to any oil or gas development by Taiwan is likely to be governed primarily by whether this takes place as part of the island's gradual movement toward implicit or explicit provincial status within a "one China" framework or rather as part of an effort to develop the independent economic strength necessary to become a sovereign republic. These alternatives are linked, in turn, with the issue of whether Taipei continues to claim the areas embraced in the northern concessions or accepts new sea boundaries reflecting its more limited jurisdiction as an island.

In this connection, it is noteworthy that even Taiwanese opponents of the Nationalist regime who favor the establishment of a sovereign republic believe that Taipei should resile from claims to offshore areas based on its purported jurisdiction over the mainland. Ng Yuzin, a leader of the "Taiwan Independence Movement" in Tokyo who edits its monthly magazine, *Taiwan Chenglian* (Young Taiwan), declared that Taipei should accept a median line roughly coincident with the boundary between the Gulf and Oceanic concessions (Figures 5 and 11). "The government can make good use of oil concessions to keep Taiwan separate," Yuzin said, "but only if it is realistic. We should stay on our side of a line that can be accepted by both China and Japan. Our movement hopes that the American oil companies will stay in Taiwan because we need their help and the people want them to stay."

The Nationalist regime was carefully keeping its options open in mid-1977, hoping for continuing economic and diplomatic support from the United States that would strengthen its bargaining posture vis-à-vis Peking. The American oil companies in Taiwan were also moving cautiously, increasingly reluctant to invest major sums in new exploration but still convinced that there was petroleum to be found and still anxious to see the island keep its independence in some form. As their concessions expired between 1978 and 1980,[45] each of these companies would have to decide whether to seek a renewal, and Taipei would have to decide whether it was desirable to have the continuing involvement of U.S. interests in view of Taipei's evolving relationships with Washington and Peking. For all concerned, the big question mark was whether the United States would sever diplomatic relations with Taipei in order to normalize its ties with Peking and what subsidiary agreements, if any, would then emerge from the normalization process regarding the future of U.S. private investments on the island and the extent of Taiwan's offshore boundaries.

CHAPTER SIX

Offshore Oil
and the
Future of Korea

As an embattled adversary in an unresolved civil conflict, South Korea, like Taiwan, has long been peculiarly captivated by dreams of an offshore oil bonanza. For Seoul, self-sufficiency in energy resources would not only have immediate economic meaning in the South Korean context but would also contribute to its larger political, psychological, and military struggle against Pyongyang. By the same token, Pyongyang has been dedicated to frustrating South Korean oil development, especially since the most promising offshore deposits adjacent to Korea appear to be located south of the 38th parallel. Pyongyang lays claim to these deposits as part of its assertion of sovereignty over the whole of Korea, and its ally, Peking, has helped to block Seoul's offshore ambitions by refusing to negotiate a boundary agreement in the Yellow Sea with an "illegitimate, puppet government."

Even in a limited economic perspective, the importance of offshore oil discoveries would clearly be enormous for the South against a background of rapidly multiplying energy needs and costs. Crude oil imports have imposed an increasingly onerous burden on the South Korean balance of payments, with the cost of crude purchases jumping from $100 million in 1969 to $277 million in 1973 and $1.4 billion in 1975. The 1975 figure represented 19 percent of the South's import bill for that year, and partly as a result of soaring oil prices, Seoul has been plunging more and more deeply into foreign indebtedness ($6 billion in 1975) in order to support its economic expansion. In mapping development plans for the 1974–81 period, South Korean leaders saw no escape from a mounting burden of oil imports, with a 2.2-fold increase anticipated in the demand for energy over the 1973 level and an increase in the ratio of oil products to overall South Korean energy consumption from 53.3 percent in 1973 to 57.8 percent by 1981.[1]

Searching
for a Bonanza

As early as 1952 Seoul had staked its claim to the undersea mineral resources surrounding South Korea by promulgating the so-called Rhee Line, which encompassed substantial areas of the continental shelf variously extending from 60 to 200 miles offshore.[2] The Rhee Line was established primarily to keep Japanese fishermen out of South Korean waters, however, and Seoul did not begin to think in offshore oil terms until the appearance of the 1961 Emery-Niino study cited in chapter 3. In 1963 the Geological Survey of Korea attempted to launch an exploration program that was limping along, hampered by limited technical expertise and a lack of funds, when the United Nations-sponsored CCOP was launched in 1966. By May 1968, with the help of a CCOP aeromagnetic survey conducted by a U.S. Navy plane,[3] the

Geological Survey had concluded that there were "good chances for thick deposits of oil or natural gas" in parts of the continental shelf to the west, south, and east of South Korea and had completed broad-brush mapping of these areas.[4] Then came the first intensive CCOP seismic surveys of the East China and Yellow seas in the fall of 1968, resulting in the much-publicized Emery report with its optimistic judgment that the sediments beneath the continental shelf and in the Yellow Sea were believed to "have great potential as oil and gas reservoirs." [5]

The Emery survey prompted euphoric official assessments of South Korean petroleum prospects, resting primarily on confident predictions of early offshore discoveries. In December 1968 the government-sponsored Korea Oil Corporation published a study declaring that the development of offshore oil resources was expected to "go into full swing beginning in 1972." The study emphasized Emery's geological characterization of the Yellow Sea sedimentary deposits west of Kunsan and southwest of Cheju Island as Paleogene and Neogene in age: "Since about 80 percent of the total underwater oil production in the world is from such Paleogene or Neogene deposits," the study said, "there is high hope that petroleum can be developed in Korea, too. What is even more encouraging is the fact that the seismic results verified sedimentary thicknesses up to 2,000 meters." At the same time, the study stressed that offshore development costs three or four times as much as onshore development and could best be undertaken with the technical and financial participation of foreign oil companies.[6]

The progress of South Korean plans for offshore exploration had been closely watched by Western oilmen with interests in Seoul. Gulf in particular had been jockeying to be first in line for any offshore concessions granted. Gulf had established close ties with the Park Chung-hee regime following the 1961 military coup by making substantial investments that constituted a resounding vote of confidence in the future of the new government. In addition to providing an immediate outlet for some of its Kuwait crude at a time when there was a glut of oil on the world market, South Korea was on the verge of significant industrial expansion, and Gulf officials saw growing market and investment opportunities there. Gulf officials actively helped Park in mapping his economic plans and encouraged other American firms to invest in South Korea.[7] In 1963 Gulf helped to establish the semigovernmental Korea Oil Corporation and received its quid pro quo in the form of an exclusive contract for the supply of crude to Korea Oil's Ulsan refinery until 1979. This partnership was to mean progressively growing profits from crude sales as prices rose, over and above a Gulf investment stake of more than $200 million by 1976 in refining, fertilizer, and petrochemical manufacturing facilities. In Gulf's eyes, offshore oil exploration was not only an attractive gamble in itself but was also viewed as a necessary public relations gesture that would serve to protect the company's dominant position on the South Korean oil scene by helping the government to achieve a major economic development objective.

In April 1969 Gulf was awarded the first two offshore concessions granted by Seoul, both of them in Yellow Sea areas along the west coast. This was followed by Shell and Texaco concessions in January and February 1970 and six months later by a Wendell Phillips award, which was eventually transferred to a consortium known as the Korean-American Oil Company (Figure 8). Significantly, both Shell and Texaco made a point of getting diversified locations for their concessions. With one location each along the southeastern coast (in an area embracing the Korea Strait and parts of the Sea of Japan and the East China Sea), in addition to one each off the western coast (in the Yellow Sea), Shell and Texaco had greater flexibility than Gulf in adjusting their exploration plans when survey and drilling operations in the Yellow Sea were complicated by jurisdictional conflicts with China. The sea boundary legislation governing these concessions sowed the seeds of future trouble by invoking different principles of international law in the Sea of Japan and the

Korea Strait, on the one hand, and the Yellow Sea on the other.[8] In the former case, Seoul utilized the *natural-prolongation* principle, which treats the continental shelf as an extension of the continent belonging to the continental country concerned; in the latter, the concessions were based on a hypothetical median line.[9]

Gulf and China: "Forcing a Resolve"

At first, China made little effort to interfere with the seismic survey ships that crisscrossed the Yellow Sea after Seoul had granted its concessions to foreign companies. Gulf-sponsored vessels alone chalked up 7,261 miles out of a total of 9,164 miles recorded by U.S. and European geophysical companies active there during the 1969–72 period.[10] As this survey work grew more intense, however, Chinese naval craft began to harass survey vessels operating relatively far from the Korean coast in a potentially disputed middle zone of the Yellow Sea.

In Chinese eyes, Seoul had acted provocatively in allocating concessions "unilaterally" without first reaching a boundary agreement with Peking or Pyongyang or both. China does not yet accept the median-line principle in Law of the Sea discussions and could well insist on geological criteria for sea boundaries more favorable to its interests. Such criteria, in turn, could lead to substantial Chinese claims extending into what South Korea regards as its side of the Yellow Sea. More important, as noted above, Peking does not recognize South Korea as a legitimate state; and even if it did agree to negotiate a median line with Seoul, this would not automatically make it easy to agree on a boundary settlement. Median-line boundaries are fixed in accordance with the particular islands, or *base points*, designated by the countries concerned as defining their coastal limits. In the case of the Yellow Sea, Chinese maps have delineated implicit base-point claims that were ignored in the initial concession boundaries laid down by South Korea.

Beginning in 1971, China conveyed its displeasure over these boundaries by sending lightly armed fishing vessels into the vicinity of survey operations. The floating tracer cables used in seismic studies were systematically cut on at least four occasions, costing the companies involved approximately $6,000 each time. As a result, the Korean-American consortium suspended seismic surveys in May and June 1971. Later, when Gulf conducted drilling operations from February to June 1973 in one of its two concession areas, known as Zone II (Figure 8), Peking escalated its response by dispatching Komar-class gunboats. The Chinese boats appeared intermittently less than a mile from the Gulf drilling rig *Glomar IV*, remaining menacingly nearby for three days in early March. This encounter was followed by a Chinese Foreign Ministry statement on 15 March 1973, attacking the drilling activities of the *Glomar IV* as

the latest step taken by international oil monopolies in their attempt to grab China's coastal sea-bed resources. . . . The sea-bed resources along the coast of China belong to China. The areas of jurisdiction of China and her neighbors in the Yellow Sea and the East China Sea have not yet been delimited. Now the South Korean authorities have flagrantly and unilaterally brought foreign oil companies into the aforementioned region for drilling activities. The Chinese government hereby reserves all rights in connection with the possible consequences arising therefrom.

Seoul interpreted the reference to "China and her neighbors" as an indirect invitation to negotiate and responded promptly with a statement on 16 March 1973 offering to hold talks with "the authorities of the People's Republic of China on the question of the delimitation of the continental shelf areas between the two countries." But China maintained a stern silence, and six days later Gulf quietly capped the well it had been drilling and shifted its operations to a new site within the same concession zone. Finally, on 10 June 1973 Gulf terminated its drilling in Zone II. The shutdown was attributed to disappointment over the lack of a discovery, but it was evident that the threatening Chinese attitude had played a major part in discouraging further exploration there. A report by Gulf geologists prepared for a technical conference in late 1974 made it clear that the most promising parts of the zone in geological terms were the Kunsan Basin and the western Yellow Sea Subbasin, both located at the western end of the concession area where Chinese and South Korean claims appeared to overlap (Figure 3). The report stressed that the two 1973 wells had provided data on only "a limited portion of the area" and that a "considerable area remains to be tested by the drill." [11] By 1976, however, these were the only two wells that had been drilled in the Yellow Sea, although four other offshore wells had been drilled in other areas with mixed results.

The danger of a conflict over sea boundaries between Peking and Seoul has been aggravated by the fact that the most promising geological structures in the Yellow Sea are located either closer to China than to South Korea or in the shadowy middle zone that has yet to be demarcated in accordance with agreed base points. From the outset Gulf knew in general terms that the most attractive structures not only in Zone II but also in its other South Korean concession, Zone IV, were located in the western portions of the areas involved and were geologically linked with more extensive structures still closer to China. [12] Acutely sensitive to the political hazards inherent in this situation, the State Department has consistently urged the two U.S. companies involved in the Yellow Sea—Gulf and Texaco—to stay close to the South Korean coast and to avoid drilling in areas that could involve a median-line controversy with Peking. Texaco was able to fulfill its contractually required exploration obligations in South Korea within Zone V, [13] off the southeastern coast. However, Gulf, with both of its concessions located off the west coast, faced a more difficult dilemma; while Washington was urging caution, Seoul was badgering Gulf to begin drilling in Zone

IV. The Seoul rumor mill hummed with reports in late 1973, heatedly denied by Gulf, that oil had actually been found in Zone II but that Gulf had capped the well temporarily in order to avoid political complications. Largely in deference to the State Department, Gulf held off drilling in both of its zones throughout 1974. The State Department argued behind the scenes that the western boundary of Zone IV, in particular, should be moved to the east of the line originally declared by Seoul, and according to Gulf sources, the company initially agreed to revise the line on an unofficial, de facto basis without surrendering its formal rights as defined in its concession agreement. But Gulf representatives in Seoul "expressed strong irritation," as American ambassador William Porter reported in a dispatch to the State Department, "inquiring sarcastically whether the Republic of Korea has any territorial waters at all" and demanding that the U.S. government tell the company, once and for all, exactly how close to the Korean coast the "disputed territory" extends.[14]

Viewed in terms of international law, the issue at stake was whether Peking had a right to claim a group of four islands, located between thirty-eight and sixty-nine miles from the Chinese coast, as appropriate base points for a median line or whether the line should properly be drawn on the basis of three different islands closer to shore.[15] Robert D. Hodgson, Geographer of the State Department, had advised Gulf that China could be expected to press its maximum claims, citing "the relevant Chinese decree . . . and a careful research into precisely where incidents have taken place and when 'serious warnings' were given."[16] Gulf's lawyer, Northcutt Ely, a veteran defender of leading oil companies in sea boundary disputes, appeared to make a gesture to his State Department adversaries in a paper presented in August 1974, conceding that a Yellow Sea median line should recognize Chinese islands as "base points even where such islands are more than 24 miles from the mainland."[17] By October, however, Ely had revised the passage in question, explaining that the original had contained a typographical error. The new draft declared that a Yellow Sea median line should recognize islands "fairly included as Chinese base points, at least to the extent that it includes islands within 24 miles of the coast." [18]

For Gulf, the issue of just how far South Korean jurisdiction did extend in the Yellow Sea had acquired an urgent practical importance. One of its proposed drilling sites was located just within what would be the Korean side of a median line if the projected boundary were drawn on the basis of probable Chinese base-point claims. After detailed geological analyses, however, this structure had proved to be only moderately attractive when compared with another one located just beyond this proposed limit on what would clearly be the Chinese side of the line (Figure 8). On the basis of its new studies, Gulf decided to insist on its rights to the original western boundary of Zone IV, as specified in the concession agreement, and made plans to include South Korea in the itinerary of a drillship, the *Wodeco VIII*, that had been engaged to work under contract to Gulf during 1975. The rig was scheduled to drill initially in the Gulf concession north of Taiwan and then to proceed to the Yellow Sea in the late spring or summer.

In contrast to the relatively quick capitulation of Superior related in chapter 5, Gulf proved hesitant to alter its plans. During a series of meetings at the State Department between January and April 1975, Gulf argued that the South Korean concession could be upheld under international law and that China had not, in any case, clearly spelled out its boundary claims. Reinforcing this stand was the deeply held belief of some of the Gulf officials involved that it would be desirable to smoke out the Chinese attitude, thus expediting a boundary settlement and opening the way for the full exploitation of oil resources that were badly needed by China, South Korea, and the United States alike. One of these officials told me in late 1974:

No oil company will withdraw from that part of the world until oil has been found and we've forced a resolve. Not until the claims have been resolved and we've been told by the governments concerned. Finding oil would act as a catalyst, and we could get a resolve.

Since the State Department was "namby-pamby," he suggested, and there was no prospect of an early boundary agreement through the diplomatic route, it was up to private industry to move things along.

An additional factor that stiffened the Gulf stand was the impatient prodding of the South Korean government to go ahead with drilling. Gulf was not only under pressure as a result of its investments in South Korea and crude oil sales that had risen above $800 million per year by 1976, but in a more direct sense, Seoul held a whip hand over Gulf in the form of price controls on the oil products sold by Korea Oil, in which Gulf had upgraded its interest to 50 percent. When Gulf showed signs of wavering in its commitment to drill in Zone IV, Seoul threatened to prevent the conversion of Gulf's profits from Korea Oil and other ventures into dollars. Gulf then asked the State Department, in turn, whether it would pledge to make good on U.S. government investment guarantees with respect to the conversion of Gulf's Korean currency in the event that a failure to drill led to South Korean reprisals.

As its exchanges with Washington dragged on, Gulf pressed for assurances that proposed revisions of the concession boundary line under a variety of different base-point scenarios would assure an acquiescent Chinese response. But the State Department stressed that the Chinese response could not be predicted with such precision, and Gulf continued to keep its options open. Deputy Secretary of State Robert S. Ingersoll, who had known Gulf Board Chairman Bob R. Dorsey in his business days, met with Dorsey several times to underline the department's concern. Dorsey made no promises, reportedly pleading that he was under strong pressure from President Park. Gulf's attitude began to change, however, when its junior partner in the Zone IV plan, Zapata Exploration, proved more amenable to U.S. government advice. When Zapata president J. B. Harrison heard Ingersoll and Assistant Secretary of State for East Asian Affairs Philip Habib lay it on the line on 4 March 1975, he quickly concluded that Zapata should extricate itself from the Korea deal.[19] "South Korea exists practically at the will of the U.S., after all," Harrison recalled later. "So if the U.S. won't uphold your title in a place like that, it really has little value." While there were no threats, he added, "It just isn't good business to be in conflict with the U.S. government. But the geology was favorable, and we were sorry to give it up." Harrison flew directly from Washington to Pittsburgh and told Dorsey how he felt, but Gulf did not make up its mind until the uproar in May over Dorsey's revelation in Senate testimony that the company had been forced to pay $4.2 million in slush funds to Park's ruling Democratic Republican party prior to the 1971 South Korean elections. The front-page publicity resulting from this disclosure aggravated a pervasive public relations problem that Gulf faced throughout the 1974–75 period as a result of foreign and domestic political contributions. Partly to avoid calling further attention to its Korean involvement and partly to appease Washington, Gulf called off its drilling plans. In early 1976, convinced that it would never be able to drill in the desirable parts of either Zone IV or Zone II, the company told Seoul that it would take a "cooperative attitude" in negotiating a transfer of the concessions if a non-American drilling company wanted to acquire them and was ready to drill there. Seoul gradually gave up hope that any U.S. company would be willing to explore in the Yellow Sea and no longer pushed Gulf as hard as it did in earlier years. In April 1977, when both concession agreements expired, a Gulf spokesman said that the company had relinquished the two areas. "We had little choice," he said, "once we realized we couldn't drill there."

While it was unclear whether Gulf had been able to fulfill its contractual commitments,[20] Seoul did not press the issue, in any case, given Washington's position on the boundary issue. To some extent, Gulf's stake in South Korea had been diminished by increasingly tight price controls on oil products and by Kuwait's nationalization of Gulf's interest in the Kuwait Oil Company, which had reduced the profits accruing to Gulf from its sales of Kuwait crude to the Korea Oil Corporation. Nevertheless, the Korean connection was still highly profitable for Gulf, given its continued control over the marketing of Kuwait crude, and the company's stake in Seoul will still be considerable if its exclusive crude supply contract with Korea Oil is renewed after its expiration in 1979.

Seoul in the
Shadow of Peking

From time to time, South Korean officials have talked of setting up an omnibus government oil corporation to handle offshore exploration and production directly along with oil import deals with the major producers, storage, refining, and the processing of oil products.[21] Such a move could conceivably lead to government contract arrangements with foreign drilling companies and eventually even to an independent South Korean drilling capability. Like Taiwan, which has acquired experience in rig manufacturing through its tie-in with Fluor Drilling Services, South Korea has already begun to acquire know-how in rig construction through manufacturing links with foreign firms. In 1976 Maersk Drilling of Denmark and Marathon, a leading American rig company, were making offshore rigs in the Korean Shipbuilding and Engineering Company yards at Pusan. Seoul had hopes that Maersk or another West European firm would take over Gulf's concession or agree to drill on a contract from the South Korean government and budgeted $607,930 for offshore exploration in 1976–77. The Geological and Mineral Research

Institute also proudly unveiled a seismic survey vessel of its own. Oil fever ran high for months after a minor onshore strike near Pohang in December 1975, and leading geologists pointed to China's successful prospecting activities in nearby waters as evidence that South Korea, too, must have oil.[22]

When I met with Kim Chan-dong, vice-minister for commerce in charge of natural resources, he insisted that South Korea was determined to develop its Yellow Sea oil regardless of pressures from China, North Korea, or the United States and regardless of whether U.S. companies are willing to drill there. "Friendly countries may say things," he declared "but the decisions will be made by the Korean government in our own interests, regardless of pressures and what others tell us our interests are." However, the South Korean attitude toward "going it alone" in the Yellow Sea has become increasingly intertwined with diplomatic efforts to get Chinese recognition of Seoul in return for American recognition of Pyongyang. Seoul appears more than willing to soften its offshore boundary claims if there is real hope of an accommodation with Peking that would loosen its ties with Pyongyang. South Korean leaders even talk privately of withdrawing their recognition of Taiwan if this would facilitate better relations with Peking. Conversely, if Peking remains firmly wedded to unilateral ties with Pyongyang, sentiment would grow in Seoul for unilateral action in the offshore oil field, especially if South Korean ties with Washington should diminish in future years.

In addition to the Yellow Sea, another possible locale for unilateral South Korean moves could be the area covered by a controversial Seoul–Tokyo treaty for joint offshore development (Figure 7) if difficulties develop over its implementation. Initialed in 1974 and routinely ratified by South Korea, the treaty represented an attempt to defuse overlapping South Korean and Japanese offshore claims by creating a joint commission to organize exploration and production activities and to implement a yet-to-be-devised formula for sharing costs and revenues. The agreement did not seek to resolve the issue of sovereignty over the areas involved, carefully side-stepping the basic dichotomy between Japanese adherence to the median-line concept and South Korean insistence on continental shelf rights under the natural-prolongation principle. In a crucial passage, the treaty stressed that "nothing in the agreement shall be regarded as determining the question of sovereign rights over all or any portion of the joint development zone, or as prejudicing the position of the respective parties with respect to the delimitation of the continental shelf." This clause gave the agreement an element of inherent instability, for it meant that both sides could still revert to the conflicting legal principles on which their original concession boundaries had been based in the event of disagreements over its implementation. Seoul's claims to part of its Zone V and all of its Zone VII had been based on the premise that its shelf jurisdiction extended up to a deep undersea trough marking a clear geological separation from Japan's own shelf. Japan had demarcated its own Zone III in much of the same area, utilizing the median-line principle.[23]

Soon after the conclusion of the agreement, North Korea cast another cloud of uncertainty over its future by declaring that the agreement was

not only totally damaging to the Korean people but also violates the autonomy and vested interests of our country. The People's Democratic Republic of Korea refuses to acknowledge this "Agreement" and declares it invalid. The continental shelf of our country is a precious treasure for the prosperity of our people, an inviolable treasure, and yet in spite of this, Park Chung-hee and his group, who would do anything for their minority interests, have not only given rights over the continental shelf to American and British imperialists but have now sold off the continental shelf in the southern sea to the Japanese militarists.[24]

China then charged that the agreement had been concluded "behind China's back," stating:

This act is an infringement on China's sovereignty, which the Chinese government absolutely cannot accept. If the Japanese government and the south Korean authorities arbitrarily carry out development activities in this area, they must bear full responsibility for all the consequences arising therefrom. . . . The Chinese government holds that, according to the principle that the continental shelf is the natural extension of the continent, it stands to reason that the question of how to divide the continental shelf in the East China Sea should be decided by China and the other countries concerned through consultations.[25]

In actuality, the greater portion of the area encompassed in the treaty was beyond the scope of probable Chinese claims,[26] but the impact of the Chinese blast was nonetheless powerful in Japan. For three years, the treaty was hopelessly ensnarled in a tangle of Japanese political infighting that reflected not only the bilateral boundary conflict between Tokyo and Seoul but also the larger political interplay between Japan, China, North Korea, and South Korea. When the Japanese government made its first futile attempt to win Diet ratification in early 1974, the treaty offered a target for opposition groups and Peking-inclined elements in the Liberal Democratic party (LDP) who were angered by the refusal of rightist LDP factions to ratify a Tokyo-Peking air agreement. In 1975 the agreement was still an intra-LDP political football, but the issues changed, with Seoul-inclined rightists demanding its ratification as a condition for approval of a proposed Sino-Japanese friendship treaty and of the long-stalled nuclear nonproliferation treaty. Much more than mere factional jockeying was involved, however, and the long debate over the joint accord took on an increasingly substantive character.

Proponents of ratification focused upon the need for oil sources close to home as an answer to OPEC price increases and confidently pointed to the prospects in the joint development zone. One enthusiastic account hailed "highly promising" geological studies said to confirm the existence of a massive, twenty-five-mile-long structure in the southwest corner of the joint zone and estimated that the zone as a whole had petroleum reserves ranging from 5 to 10 billion tons (37.5 to 75 billion barrels).[27]

Leading newspapers countered that the median-line principle was likely to be adopted by the United Nations Law of the Sea Conference and that Japan could claim the entire zone for itself by invoking this principle. Equally important, it was argued, Japan would weaken its case in rebutting Chinese continental shelf claims if it failed to contest South Korean claims based on the same natural-prolongation principle used by Peking.[28] The fisheries lobby joined in the clamor against the treaty by warning against the pollution likely to result from offshore oil development. Above all, *Mainichi* warned, Japan would be inviting "serious future political trouble" with China and North Korea alike by siding with South Korea. The Park regime was inherently unstable and unpredictable, the paper said, citing the abduction of South Korean opposition leader Kim Dae-jung from a Tokyo hotel room "in violation of Japanese sovereignty." The Chinese statement attacking the agreement with Seoul marked the first serious criticism of Japan to emanate from Peking since the 1972 normalization of relations, *Mainichi* noted, and adherence to the joint accord might even jeopardize the negotiations for a friendship treaty with Peking.[29] Given the many obvious reasons for exercising caution, *Asahi* observed, it was "puzzling" that the agreement had been concluded "so hurriedly, as if evading the eyes of the people, with no consideration of the national interest either by the Cabinet or the LDP. This makes us feel that it is beclouded by the shadow of unwholesome vested interests." [30]

The suspicions expressed by *Asahi* are widely shared in Japan and imparted a note of unusual bitterness to the protracted treaty debate from 1975 to 1977. As chapter 7 spells out, active support for the treaty in Japan was centered in business circles with ties to Seoul, some of them with a direct interest in offshore oil, or in kindred LDP rightist factions frankly dedicated to maximizing relations with South Korea as against North Korea. Most other segments of Japanese opinion were at best lukewarm, except for some of the more ardent advocates of energy independence, and the treaty was finally ratified in June 1977 only as the by-product of unrelated developments affecting politically powerful Japanese fishing interests. Just as pro-Seoul elements had sought to link ratification of the Seoul–Tokyo treaty with the nuclear nonproliferation issue in 1974 and 1975, so they attempted to use the ratification of a long-stalled fishing agreement with the Soviet Union to neutralize opposition to the Seoul–Tokyo accord in 1977. This time they were successful, employing a complex parliamentary maneuver in the Japanese Diet to make the fate of the Soviet and South Korean accords interdependent. Many deputies who were opposed to the Seoul–Tokyo treaty were forced to go along with it in order to register their support for the agreement with Moscow, a matter of vital economic concern to the fishing lobby.

South Korean leaders are closely attuned to the Japanese political scene and have been hopeful that the lure of oil and oil-related profits will help to strengthen their allies in Tokyo, thus furthering a Seoul-Tokyo alignment at the expense of Pyongyang. In their more optimistic moments, South Korean officials have even expressed the belief that a Japanese government dominated by their allies might eventually be able to induce China to enter into sea boundary talks with Tokyo and Seoul, setting the stage for a tripartite renegotiation of the treaty that would isolate Pyongyang and give new legitimacy to South Korea. As a consequence of the long delay in ratification, however, Seoul has grown skeptical of Japanese intentions and has talked more and more openly of asserting its claims unilaterally in the event that Tokyo should fail to implement the treaty.[31] The consistently defensive Japanese government stance on the treaty in the Diet [32] and the precarious circumstances surrounding its ratification have prompted Seoul to question whether Tokyo will ever actually pass the necessary follow-up legislation. Until agreement has been reached on the operational modalities of oil development, many South Koreans are likely to view the treaty as a cynical Japanese ruse that was intended, from the start, to tie Seoul's hands and buy time for Tokyo in its dealings with Peking and Pyongyang.

Whether or not such suspicions are valid, Japan was clearly shaken by the vehemence of the Chinese reaction to ratification of the agreement. In a formal protest, Peking went far beyond its 1974 statement, which had cited the natural-prolongation principle in support of China's right to a voice in delimiting the continental shelf but had stopped short of explicitly claiming the shelf. This time, the Foreign Ministry flatly stated on 13 June 1977 that:

According to the principle that the continental shelf is the natural extension of the continental territory, the People's Republic of China has inviolable sovereignty over the East China Sea continental shelf. . . . Certain members of the Japanese government circles even alleged that "the Japan-South Korea joint development zone is restricted to the Japanese side of the intermediate line of equal distance between Japan and China," and does not infringe on China's sovereignty. Such an argument is futile and utterly untenable. . . . Without the consent of the Chinese government, no country or private individual may undertake development activities on the East China Sea continental shelf, and whoever does so must bear full responsibility for all the consequences arising therefrom.

As in 1974, the statement said that "the question of how to divide those parts of the East China Sea which involve other countries should be decided by China and the countries concerned through consultations." But Peking's bargaining posture had been substantially stiffened by its claim that the shelf "forms an integral part of the mainland."

With Sino-Japanese trade expanding, Tokyo's stake in good relations with Peking was growing in mid-1977. As chapter 7 shows, Japan wanted to enlarge its oil purchases, in particular, drawing on both onshore and offshore deposits at Taching, Takang, Shengli and the Po Hai Gulf that are located within temptingly easy tanker reach. The promise of progressively growing oil imports assures a continuing Japanese interest in ties with Peking and has led Japan to soft-pedal its continental shelf claims. As long as the Chinese oil tap continues to flow, Japan appears unlikely to press its claims vis-à-vis China in oil-rich areas on the shelf north of Taiwan and will no doubt move cautiously in implementing the joint development agreement with Seoul. Moreover, as an offshoot of its growing intimacy with Peking, Tokyo is generally more wary of becoming overcommitted to Seoul than in earlier years and is instead seeking to balance its South Korean involvements with new links to Pyongyang.

The strong Chinese commitment to Pyongyang in the North-South struggle has a powerful impact on the triangular Japanese-Chinese-Korean relationship that is often underemphasized. Peking must give top priority to its Pyongyang ties as a critical factor in its rivalry with Moscow, making it necessary for Peking to support Pyongyang unambiguously in its effort to nullify both the Yellow Sea concessions granted by Seoul and the Japan-South Korean agreement. Japan, in turn, must weigh the immediate advantages likely to result from a dependable oil partnership with Peking. Looking ahead, a Tokyo-Seoul-Peking agreement on offshore oil might have practical economic advantages for all concerned but is most improbable in the light of political realities. Unless Peking should decide to wink at joint action by Seoul and Tokyo, it appears likely that significant development activity will be paralyzed in the joint zone unless and until tensions ease in the Korean peninsula and new arrangements, such as those discussed in chapter 11, can be made by Japan and China with the concurrence of both Seoul and Pyongyang. It should be noted here that China carefully used the word *development* in warning Japan and South Korea not to "arbitrarily carry out development activities in this area," implying that seismic studies and other exploration, short of development, might not evoke Chinese reprisals. Conceivably, exploratory studies of the joint zone by Japan, South Korea, and possibly China as well might take place in the interval preceding an agreement on development.

Given the proximity of the Korean peninsula to the major centers of Chinese oil production, it is possible that both North and South Korea, like Japan, will look to China not only for more of their oil imports but for cooperative relationships in offshore oil development. Already, Pyongyang is leaning heavily on Peking for its oil supplies, importing more than 1.2 million tons (9 million barrels) from China in 1974 as against an estimated 585,000 tons (4.4 million barrels) from the Soviet Union, a sharp reversal of its earlier dependence on Moscow. As industrialization has progressed in the North, fuel imports have increased from 8.4 percent of all imports in 1956 to 19.3 percent in 1969, and a comparable increase appears to be steadily continuing.[33] Ironically, although built with Soviet help, Pyongyang's two refineries have become more and more oriented to Chinese supplies as Peking's export capacity has grown, culminating in the announcement of a direct pipeline link from the Taching oil field to North Korea in January 1976. So far, as observed earlier in this chapter, none of the available evidence suggests the existence of large-scale offshore deposits adjacent to the North, nor is there any indication that Pyongyang has ambitions for a significant offshore program of its own. Should the North seek to develop offshore oil, however, it is noteworthy that Pyongyang announced its acceptance of the median-line principle at the Caracas Law of the Sea Conference in the summer of 1974.

Despite its political isolation from Peking, Seoul has been acutely sensitive to the massive expansion of Chinese petroleum production taking place just across the Yellow Sea. China's emergence as a major oil and gas producer has come at a time when Seoul is deeply resentful of the steady rise in the price of its imported crude and disenchanted with the reluctance of American firms to drill in offshore areas contested by Peking. By opening up an oil pipeline to the North in 1976, Peking has given Pyongyang new staying power in its immediate confrontation with Seoul. In long-range terms, the new pipeline has also dramatized the fact that the Korean peninsula as a whole lives under the shadow of an increasingly powerful neighbor and that the North-South political impasse prevents the South from balancing its dependence on the Middle East with offsetting energy supply links to Peking. Appealing for a more "subtle diplomacy" designed to bring Peking to the conference table for a discussion of the Korean issue, opposition leader Kim Young-sam pointedly emphasized that China is "a vast country with a huge population, powerful military strength and abundant natural resources." [34] It is not inconceivable that backdoor arrangements will evolve in future decades for Chinese oil exports to South Korea, routed through third parties, or even for Chinese assistance to Yellow Sea oil development by South and North Korea alike. As Peking's offshore know-how grows in the Po Hai–Yellow Sea area, so will its bargaining power in facing Pyongyang and Seoul or the leaders of some future unified Korea, as well as its desire to have a decisive voice in determining how the Yellow Sea is to be parceled out and how its development is to be organized. The expansion of Chinese offshore activity could lead to significant areas of beneficial regional cooperation, but it could also result in serious boundary conflicts arising from the geological interdependence of the Yellow Sea, discussed earlier, and the possible difficulty of finding an agreed formula for dividing it up. Peking's seizure of two South Korean fishing boats in early 1976—the fifth such incident in a decade—was a reminder of Chinese sensitivities.

At the very least, renewed drilling by Seoul prior to a boundary agreement, either on its own or in cooperation with foreign companies, could provoke a Chinese political reaction that would aggravate the already delicate interplay between Peking, Pyongyang, and Washington. At worst, it could trigger a military explosion should it coincide with an overall escalation of tension in Korea.

CHAPTER SEVEN

China, Japan, and the Oil Weapon

At first glance, the most serious danger of offshore oil conflict between China and its neighbors appears to lie in the possibility of a Sino-Japanese clash over the division of the East China Sea continental shelf. Peking has asserted implicit claims to the entire shelf, as we have seen, and has not yet accepted the median-line principle as a valid basis for resolving sea boundary disputes. Tokyo, for its part, envisages a median line that would place the most promising areas for oil exploration under Japanese jurisdiction, invoking its claim to the disputed Senkaku Islands (Tiao-yü T'ai) in support of its boundary demands. All the elements necessary for an explosion appear to be present, with both countries gradually building up their naval power and Japan increasingly anxious to reduce its dependence on the Middle East. Taking a larger view, however, it is apparent that the East China Sea conflict has been temporarily defused by the overall pattern of improvement in Sino-Japanese relations. China has soft-pedaled its claims to the Senkakus (Tiao-yü T'ai) as part of a broad effort to offset Soviet-Japanese ties and has induced Japan to pigeonhole its own claims, so far, by offering the bait of long-term oil export commitments. Japan is cautiously measuring the potential benefits of an oil-centered trade partnership with China against the costs of pressing its claims in the East China Sea and of undertaking offshore exploration on its own. By avoiding a collision with China, Japan sees a possible opportunity to obtain the lion's share of offshore oil and gas production in East Asia without incurring the attendant exploration risks. At the same time, many Japanese leaders hope for a lucrative technical partnership role in the development of Chinese offshore resources, first in the Po Hai–Yellow Sea area and later in the East China Sea as well.

What Chinese Oil
Means to Japan

Political and economic factors combine and overlap to make the prospect of large-scale oil imports from China peculiarly alluring for Japan. At one stroke, the China option offers a way of offsetting the strategic vulnerability inherent in a geographically one-sided dependence on the Middle East, while gaining increased leverage in bargaining for lower prices with the producer countries and the majors.

In 1974 Japan was dependent on imports for 88 percent of its oil and dependent on a single geographical area, the Middle East, for 77 percent of this imported oil. To compound the problem, 65.1 percent of its imports were controlled by the Western majors, including oil coming from Indonesia and other non-Middle East sources. American-based companies supplied 49.5 percent, with Caltex alone accounting for 15.5 percent.[1]

Initially, when the Western majors moved in along

with the American occupation forces, Japan had slipped gracefully into what was a comfortable and highly advantageous dependence. In converting the coal-based Japanese economy overnight into one based almost wholly on oil, the majors helped to build a network of up-to-date refineries in Japan and bore nearly all of the risks involved in the rapid postwar development of the Middle East oil fields. By design, Japan became one of the principal beneficiaries of the cold war collaboration that developed between Washington and the "Seven Sisters." Dependence on the Middle East in this context meant a reliable supply of inexpensive oil that proved to be a critical factor in Japanese reconstruction and expansion. With the appearance of OPEC, however, the meaning of dependence changed, and Japan quickly perceived that its interests were increasingly divergent from those of the majors. As early as 1971, Japan was quietly belaboring the majors for passing on price increases and reaping higher profits at the expense of consumer countries. Now the dominant Japanese view is that the majors, unable to defeat the cartel, have for all practical purposes joined it and have a stake in higher price levels as a consequence of their new managerial and marketing role.

When Secretary Kissinger made his proposal for a common front of consumer countries in November 1974, *Nihon Keizai,* the Japanese counterpart of the *Wall Street Journal,* observed:

The oil producer nations have clearly connected the crude oil price reduction issue with the excess profits earned by the majors. Consumer nations will not be convinced if the U.S., which has five majors at home, limits itself to advocating the necessity of cooperation among consumer nations without touching upon the excess profits earned by the majors.[2]

James C. Abegglen has observed:

The oil crisis reminds Japan again of its vulnerability to the behavior and decisions of the international oil companies, and it is unreasonable to expect that Japan will tolerate that vulnerability over the long run. The oil crisis also reminds Japan of the fact that its interests and those of the U.S. are with increasing frequency not in parallel and . . . thus seems likely to move Japan toward disassociation from the U.S., toward direct dealing in the oil it requires, and toward involvement with the countries of the Third World that have resources and seek industry.[3]

Ideally, Japan would like to bypass the majors and establish direct deals with nationalized oil enterprises in the Middle East and elsewhere within a relatively short span of years. This prospect has been forestalled, however, by the success of the majors in retaining a significant measure of control over the marketing and distribution of Middle East oil. In 1971 Japanese refiners began to import some of their crude through Japanese trading companies rather than directly through the majors, only to find that the trading companies themselves were often forced to work through the majors. Faced with this situation, Japan has been forced to move with caution and circumspection. Nevertheless, as opportunities arise, Japan is shifting some of the imports handled by Japanese trading companies to direct deals. Over time, as contracts with the majors come up for renewal, some of these are also likely to be shifted over to purchases from national oil companies, even though such purchases might not provide an immediate price advantage and, in some circumstances, might even involve higher prices. Increasingly, one finds a bitter recognition in Japan that the diversification of energy sources in geographical terms will not be meaningful without greater independence from the majors,[4] and it is this recognition that has made the China option so attractive. For in addition to the advantages offered by geographical proximity, Peking provides a potentially powerful source of bargaining leverage vis-à-vis the majors as a result of its autarkic, free-wheeling position in the world economy.

The potential for an oil-centered trade partnership would appear to be much greater in the case of Japan and China than in that of Japan and the Soviet Union. In the

Chinese case, Japan is critically influenced by a positive psychological chemistry that contrasts markedly with its historic distrust of the Soviet Union. The prospect of large-scale oil imports from China provides a pragmatic rationale for a posture toward China that many Japanese would like to adopt anyway, apart from economic considerations. It is not possible here to do more than sketch the outlines of the complex factors involved in the Sino-Japanese relationship.[5] In essence, the view underlying this analysis is that the pro-Chinese political undertow in Japan makes it increasingly difficult for Japanese policymakers to preserve a position of equidistance between Peking and Moscow. The desire for symmetry is in continuing conflict with the atavistic pull of common ethnic and cultural roots summed up in the Japanese expression *dobun dosyu* ("same race, same letters"). Racial kinship would appear to be only one element of a stronger sense of communality that has yet to be adequately explored or explained in Western social science research. In structural terms, the Chinese and Japanese languages have diverged significantly over the centuries, but in emotional terms, the common use of Chinese ideographs appears to be an elemental common denominator. Beyond this, the bonds of culture and ethnicity are reinforced by a sense of shared destiny in the face of the West that dates back more than a century. Japan's subsequent attempt to dominate China left a legacy of fratricidal bitterness that aggravated the ideological differences dividing Peking and Tokyo after the Communist victory. By the time the two countries had re-established top-level contacts in 1972, however, they were both prepared to wipe the slate clean. In effect, they decided to go back to the turn of the century and to begin all over again. China was able to submerge its war-born animosity toward Japan in its fears of the Soviet Union, and Japan was anxiously casting about for new moorings at a time of international economic and political flux. Most important, Tokyo was compelled to come to terms with the settled fact of a strong, centralized China that was no longer vulnerable to foreign domination and had its own claims to a place of regional primacy in which Japan would be subordinate or, at best, coequal. Confronted with this reality, Japan has been sliding into a progressively more adaptive stance. A close and compatible relationship with China appears to be emotionally necessary for many Japanese, providing a sense of anchorage and security amid shifting international currents. This will not necessarily deter Japan from seeking to drive a hard bargain in its economic and political disputes with China, and a long process of adjustment no doubt lies ahead. But viewed in its totality, the overall record since the 1972 normalization suggests that Japan is more likely to be adaptive than competitive as Peking grows in relative strength and power.

The psychological and cultural factors conditioning Sino-Japanese and Soviet-Japanese relations have been clearly discernible during the triangular bargaining over energy supply arrangements between Tokyo, Peking, and Moscow that began in 1972 and has not yet fully run its course. Year after year, Japanese and Chinese delegations have patiently carried forward a protracted and exacting but generally compatible bargaining process notable for a minimal level of mutual recriminations. Conversations with many of the Japanese business and governmental leaders involved indicate that they feel relatively relaxed in dealing with their Chinese counterparts and are prepared to give them the benefit of the doubt when negotiations drag on over long periods, as in the case of the long-term oil import agreement that was under discussion in 1977. By contrast, Japanese dealings with Soviet officials over the abortive Tumen oil deal were brittle, tense, and marked by steadily growing distrust.

In light of their history of military conflict with Russia, many Japanese were disturbed by the prospect of a major energy dependence on the Soviet Union in the first place and were reluctant to give Moscow the benefit of the doubt when the Soviet negotiating stance was reversed midway through the Tumen negotiations. As originally formulated, the Soviet proposal envisaged Japanese credits of $2 billion for the construction of a pipeline from the Tumen oil field in western Siberia to the Soviet Pacific port of Nakhodka, with an estimated yearly flow of 40 million tons (300 million barrels) of crude oil for shipment to Japan. However, in late 1973 Soviet negotiators changed their tune, talking of credits exceeding $3.1 billion and an oil flow of only 25 million tons (187.5 million barrels) per year. This was later followed by suggestions for a second trans-Siberian railway to transport the crude oil over part of the distance that would have been covered by the pipeline. In Chinese eyes, even the pipeline would have been provocative, partly because it was expected to run close to the Sino-Soviet border in many areas and partly because stepped-up oil supplies would have had an obvious military value to the Soviet Far Eastern military forces based in Vladivostok. The railway would have been a still more blatant challenge to Peking and would have made the new Soviet proposal unacceptable to Tokyo, even in the absence of the other changes in the terms of the Soviet offer.

The Tumen negotiations underlined the fact that Moscow does not have already discovered oil deposits located as favorably in relation to Japan as the Chinese reserves in the Po Hai Gulf–Yellow Sea area. The Tumen fields are some 3,100 miles from Japan, while the Po Hai–Yellow Sea area is only 400 miles away by tanker. Moreover, the Tumen area is considerably closer to Europe than to Japan, and the Soviet Union was under pressure in 1973 to sell some of the oil that had been intended for Tokyo to Eastern European Communist countries. This was apparently the principal reason for the Soviet volte-face with respect to the terms of its offer to Japan. When Moscow reduced its promised level of oil via the Tumen route to 25 million tons (187.5 million barrels) by 1985, Peking not only countered with its own offer of 25 million tons by 1980 but was also able to do so without asking for the large-scale credits from Japan that the Soviet pipeline and rail projects would have entailed as a result of the distance factor. The most promising Soviet oil deposits close to Japan are those in the Sea of Okhotsk near Sakhalin, where

Soviet and Japanese geologists started to explore in 1976 under an arrangement providing for $122 million in Japanese credits, much of this sum repayable in crude and only repayable if oil is found. Japan is to get half of any oil produced for ten years in addition to the oil supplied in repayment of these credits. The arrangement was significant as an example of the type of exploration partnership that Japan has been unable to obtain from China, but it remains to be seen whether these reserves rival the deposits known to exist in the Po Hai Gulf and the adjacent Takang and Shengli fields. The Yakutsk gas fields in eastern Siberia, while highly promising and more fully explored than the petroleum reserves off Sakhalin, are much farther from Japan and much harder to reach than the gas deposits that have been discovered at Takang. Japan would like major American participation in the development of Yakutsk to dilute the serious financial and political risks that would be involved, but this does not appear to be forthcoming.

Even if offshore oil were found in the Sea of Okhotsk, Japan appears likely to attach greater importance to its energy links with Peking than to any Soviet energy sources in the context of its increasing emphasis on the overall expansion of Sino-Japanese trade. The specter of recession and protectionism in the West has accelerated Japanese efforts to diversify foreign markets as well as raw material sources, and China, together with Latin America, loom larger and larger in Japanese trade calculations. Japanese markets in Southeast Asia, while expanding, have not been growing rapidly enough to compensate for the declining growth in Japanese exports to the United States and Western Europe; and China, at its present stage of development, is viewed as a more promising new frontier than the Soviet Union. Both stress economic autarky and are inherently less attractive as markets for consumer goods than the Western countries. For the next generation, however, Japan sees more hope for a major expansion of its exports in a country just getting started in its industrial development than in one already well along in the process. An energy partnership becomes critical in the case of China not only for its own sake but as part of a wide-ranging expansion of trade in which Peking is able to pay for its imports from Japan largely through its oil exports. Soviet two-way trade with Japan had reached an estimated $3.5 billion in 1976, but Japanese exports were financed mainly by a rising volume of Japanese credits.

Long before the normalization of relations with Peking, Japanese two-way trade with China had jumped from $265.7 million in 1965 to $822.7 million in 1970 and $1.1 billion in 1972. The 1972 normalization triggered successive leaps to $2.013 billion in 1973, $3.289 billion in 1974, and $3.8 billion in 1975. However, the annual rate of increase dropped markedly from 83 percent in 1973 and 63 percent in 1974 to 15 percent in 1975. In 1976, two-way trade fell by 20.2 percent, a development that stirred profound anxieties among Japanese advocates of greater trade with China. The slowdown in the rate of growth was attributed in Japan partly to the impact of the global recession on the Japanese demand for Chinese products and partly to factional differences in Peking over foreign trade. But it was viewed, above all, as an indication that Chinese leaders of all factions were determined to place limits on a steadily widening trade imbalance in favor of Japan. As early as 1974, Foreign Trade Minister Li Chiang had warned that the further expansion of Japanese exports to China would depend upon the ability of the two countries to evolve a more balanced trade relationship. In that one year, Japanese exports to China had shot up by 91 percent, and the trade surplus in favor of Japan had reached $683.2 million even though Japanese imports from China had also risen by 34 percent.

The importance of oil in the overall pattern of Sino-Japanese trade was dramatized by the rapid rise in Chinese oil exports from 4.3 percent of the total value of Chinese exports to Japan during 1973 to 32.9 percent in 1974 and 41.3 percent in 1976. Oil has given China a way to cushion the impact of the global recession, which has pushed up the prices of Japanese exports while forcing China to restrain its prices on most items for competitive reasons.[6] Looking ahead, a progressively rising volume of oil exports would clearly be necessary to sustain a continuing growth in Japanese exports. Peking is pressing Japan to increase its imports of a variety of Chinese products,[7] but is placing major emphasis on accelerated oil imports as the key to the large-scale expansion of Japanese industrial exports.

The potential beneficiaries of expanded Sino-Japanese trade—and thus the most active proponents of increased oil imports from China—include the powerful machinery, steel, and construction industries as well as oil-related industries seeking to link up directly with Chinese oil development. Machinery and equipment exports jumped from 17.9 percent of total Japanese exports to China in 1973 to 28 percent in 1974 and 31 percent in the first nine months of 1975. In terms of value, this meant an increase from $186 million in 1973 to $555 million in 1974 and $697 million in 1975, including exports of $220.3 million in automobiles, trucks, and other transport equipment. Iron and steel exports rose to $727.8 million in value in 1974 and $824 million in 1976, and a major item in 1975 and 1976 was seamless steel pipes destined primarily for use in oil pipelines, which totaled $297 million for the two-year period. The share of iron and steel in Japan's total exports to China fell from 48.9 percent in 1973 to 35 percent in 1975 but jumped back to 49.6 percent in 1976. (The 1974–75 slump was attributed primarily to high prices and scarcities in Japan during 1974; since then a decline in the world demand for iron and steel has led to a more accommodating Japanese posture in negotiations regarding iron and steel prices.) It is no accident, therefore, that one of the most vigorous advocates of a long-term oil agreement with China has been Yoshihiro Inayama, president of Nippon Steel, who has led several trade missions to Peking and has sought to establish the concept of a direct linkage between the level of Chinese steel purchases from Japan and the level of Japanese oil imports from China.

Despite the American lead in steel technology, Japan is well situated to assume a leading role, not only in supplying steel products, but also in helping China to develop its own steel industry. The new Chinese steel complex under construction near Wuhan includes a $231-million hot-steel rolling mill purchased on a "turn-key" basis from a Japanese consortium led by Nippon Steel. Before its completion in early 1977, 350 Japanese technicians were scheduled to visit China, and 200 Chinese technicians were scheduled for training in Japan. Similarly, there is a great potential for the export of complete plants in other spheres of industry, especially refineries and other plants that would draw on Chinese oil as a raw material. In addition to the Wuhan steel complex, Japan concluded agreements with China between 1972 and 1974 for the sale of seventeen plants with a value of $470 million, among them fertilizer and petrochemical factories. A consortium of Japanese companies led by Bridgestone has been negotiating with China for the construction of a liquefaction plant near a gas field in the Takang area capable of producing 300,000 tons of liquefied natural gas (LNG) per year for export to Japan. The projected long-term oil import agreement under discussion in early 1977 envisaged a companion agreement for $1.2 billion in plant exports backed by Japanese Export–Import Bank credits.

Reflecting the impatient mood of China-minded industrialists, a leading commentator in *Nihon Keizai* warned in November 1975:

Until we correct the imbalance in our trade, the planned economy of China will have a hard time incorporating the export of Chinese resources to Japan and of Japanese exports to China into their long-range plans. Therefore, it is imperative that we resolve the continued imbalance if we are to launch large-scale project exports to China. . . . It is clear that many industries stand to benefit from the expanded import of Chinese crude oil, such as the iron and steel industry, the machinery and equipment industry, the plant construction industry and the systems engineering sector. This benefit can be realized if the oil trade with Japan becomes more lucrative for China. What is at stake in diversifying our sources of energy supplies in this fashion is nothing less than the broad functioning and welfare of the Japanese economy as a whole.[8]

Bargaining, Asian Style

Considering the power of the business interests that have a stake in expanded trade with China, there appears to be little doubt that Japanese imports of Chinese oil will grow substantially. In negotiating the terms of long-range import arrangements, however, China and Japan have encountered a variety of difficulties, among them significant technical problems resulting from the character of the Chinese oil initially exported to Japan.

Japanese refiners were quick to complain that the variety of crude exported to Japan from the Taching field between 1972 and 1975 has a higher wax content than the Arabian variety to which Japanese refineries have hitherto been geared. In addition to complicating storage and transportation, a high wax content means that the refinery distillation process results in less gasoline, kerosene, and naphtha than that yielded by Arabian crude and a proportionately greater component of residual "heavy" oil, usable only for burning in electric power plants. For example, one study showed that Taching crude yielded 12.7 percent gasoline as against 19 percent from Arabian, 7.3 percent kerosene as against 12 percent from Arabian, and 9.4 percent naphtha as against 24 percent from Arabian, with a net result of a 70.5 percent heavy-oil distillate as against 44 percent in the case of Arabian crude.[9] It is necessary to reduce the wax content by means of a costly catalytic "cracking" process in order to refine Taching crude at all, but many Japanese refineries were designed by the majors to handle Middle Eastern crudes and do not have these facilities. As a consequence, more than 50 percent of the Chinese crude imported by Japan has not been refined and has been burned directly by electric power plants. To complicate matters further, there has been a limited demand for Taching crude and Taching heavy-oil residue in Japanese power plants, which are already adequately supplied by Minas crude from Indonesia. Minas, too, has a relatively high heavy residue of 64.2 percent, and it is only with considerable difficulty that an annual market for 21 million tons (157.5 million barrels) of the Indonesian crude has been built up in Japan. As it happens, the production and distribution of Minas are controlled by Caltex, so that the substitution of Taching for Minas in electric power plants would bring a collision not only with the Indonesian government but also with a powerful Western major. Less than 10 percent of the Taching residue in the hands of Japanese refiners was actually used in power plants during the 1974 fiscal year and only 32 percent in 1975. This was a major blow for the refiners involved, especially after the costly outlays for desulfurization during the years immediately preceding the inauguration of Taching crude imports. A mood of general uncertainty regarding the future level of demand for oil added to the hesitation of the refiners when the Ministry of International Trade and Industry (MITI) called on the industry in 1975 to install cracking facilities. To ease the immediate situation, MITI was finally able to induce Maruzen and Idemitsu Kosan to make the necessary conversion for Taching crude at their Bungotakata and Hyogo refineries by arranging governmental credits. But this was far from sufficient to accommodate a large-scale expansion of imports from China, and the "wax problem" was left to be resolved as part of a comprehensive pattern of compromise between Japan and China with respect to the terms of their oil partnership.

In mid-1977 a complex tripartite bargaining process between Tokyo, Peking, and the majors was gradually getting under way. Japan was seeking to push the price of Chinese oil down as far as the traffic would bear. At the very least,

Tokyo wanted to make Peking share in the costs of the conversion of heavy to "light" oil, either through its construction of new Chinese refineries designed with the Japanese market in mind, possibly backed by Japanese credits, or through enough of a reduction in price to offset the installation of conversion facilities in Japanese refineries. Uncertain as to how great the Japanese demand for oil would be, Tokyo was resisting Chinese pressures for firm commitments to specified annual increases in imports pending concessions on the price and refinery issues. The Western majors were seeking to dissuade Japan from risking large-scale, long-term engagements with Peking at the possible expense of existing arrangements for the supply of Middle Eastern and Indonesian crude, partly by the use of gentle pressures and partly by casting doubt on the ability of China to deliver on its commitments. To cover their flanks, the majors were also exploring the possibility of purchasing crude from China for resale to Japan and other Asian customers as well as the U.S. West Coast. China, meanwhile, was seeking to utilize negotiations with the Western majors as a means of reminding Japan that Peking has other options and does not have to depend on the Japanese market. When former foreign minister Aiichiro Fujiyama visited Peking in December 1975, Vice Minister of Foreign Trade Yao I-lin pointedly mentioned that negotiations for crude sales exports to unspecified U.S. firms were at "an advanced stage," prompting an *Asahi* correspondent to comment that "this speculation appeared designed to push Japan to a decision with respect to new refining facilities and to expand the 'oil path' to Japan as the biggest market for Chinese oil in the future." [10]

In political terms this bargaining process was closely linked with the protracted jousting taking place between China and Japan over the terms of their long-stalled friendship treaty. China was seeking Japanese acceptance of an implicitly anti-Soviet clause pledging joint Sino-Japanese opposition to "hegemony" by any power in Asia, and Japan was searching for a way to avoid alienating Moscow unnecessarily. In Japan's eyes, acceptance of the hegemony clause involved major political costs threatening its relations with the Soviet Union, and Tokyo was hoping to obtain a Chinese quid pro quo in the form of advantageous oil arrangements. Similarly, Tokyo was attempting to get a Chinese go-ahead for implementation of the Japanese offshore oil agreement with South Korea as part of an overall bargain including conclusion of the friendship treaty.

In the short run, it appeared possible that limited Chinese sales to the majors could prove convenient as an interim device for China and Japan alike, at least pending the evolution of their bargaining over prices and refinery plans. Japan has been reluctant to make firm advance commitments for an increase in imports above an 18-million-ton (135-million-barrel) level covering the 1976–85 period but would like to have the option of buying more, while China, with its planned economy, wants to know where it stands and has been unwilling to give Japan this flexibility. By selling small amounts of oil to the majors for resale to Japan or other countries as the market dictates, Peking could kill several birds with one stone, adding to its foreign exchange earnings, helping to meet Japanese needs, and making a token political gesture to the United States. In a similar vein, Japanese refineries affiliated with the majors could buy Chinese oil, as some are already beginning to do, channeling any amounts that prove to be surplus to other markets through their Western partners. For China, such arrangements would not pose the fundamental political issues that would be involved in large-scale crude agreements with the Western majors. In the long run, however, Japan appears most unlikely to get into the position of relying on the majors for the bulk of its Chinese oil imports. The Ministry of International Trade and Industry is explicitly seeking to avoid this by pursuing a long-term, government-to-government agreement. The economic and technological power wielded by the majors is increasingly resented in Japan, and the majors' 1976 negotiations with Peking served to sensitize popular fears that Western interests are seeking to muscle into China ahead of Japan.

Japanese business magazines have pointed with special bitterness to the interest shown by the majors in the offshore riches of "our" continental shelf.[11] Hailing collaboration with China as a symbol of "our resurgence, our Asian resurgence," Yoshihito Shimada, president emeritus of the semigovernmental Oil Development Foundation, observed in a roundtable discussion that "when China does become an oil country, it will change the appearance of Asia. In terms of the global power balance, Asia is thought of as limited in oil resources, but that will change from now on, won't it?"[12] A *Nihon Keizai* assessment of Japan's oil negotiations with China deplored the veto power over import sources exercised by Caltex and other refineries affiliated with the majors, concluding that "we must discover how to create the framework for A New Order for Asian Oil."[13]

It could prove extremely shortsighted to dismiss the potential of the Sino-Japanese relationship on the basis of the relatively modest import levels projected in the 1976 negotiations—10 million tons (75 million barrels) per year initially, rising to 18 million tons (135 million barrels) by 1985. Most indications suggest that these levels merely represent minimum Japanese commitments and might well be upgraded as the bargaining over prices and refineries proceeds in the years ahead. Shuichi Matsune, chairman of the energy-policy committee of Keidanren, Japan's leading business federation, has estimated that a large-scale refinery conversion is likely to be completed

within four or five years, if the Japanese government goes in on it, and Japan will then be able to handle Chinese oil if it really starts coming in. [Japan] has no choice but to gear to using Chinese oil, and once this process is in motion, we will be able to meet one-third of our oil needs with oil from Asia! There will be no need to go through the Straits of Malacca. This would mean security for Japan both economically and politically.[14]

How far is Japan prepared to go in offsetting its Middle East dependence with a new dependence on Peking?

In the first flush of enthusiasm accompanying the 1972

normalization, Keisuke Idemitsu, one of the key figures in the Japanese oil industry, declared that Japan eventually hoped to import 50 percent of its crude oil and refined products from China.[15] Subsequently, MITI has suggested a target of 20 percent as the ultimate result of gradual increases in imports under a long-term agreement with Peking,[16] although some officials and business leaders, like Matsune, have talked of a 35 percent reliance.[17] More cautious observers outside the government envisage a maximum dependence of 11.6 percent by 1985.[18] Even if oil imports remain near the 300-million-ton (2.3-billion-barrel) level until 1985, imports of 18 million tons (135 million barrels) would represent only 6 percent of the total. Given the atavistic attraction to China discussed earlier, it would not be difficult for Japan to absorb a 35 to 50 percent dependence in psychological terms, but one would be rash indeed to predict such an eventuality within any given time span. China's ability to export oil and its desire to do so may periodically change and Japan's future appetite for Chinese oil imports will be governed by a variety of uncertain economic and political variables.

The most important of the economic variables involved is likely to be the overall state of the world petroleum market. In a slack market, Japan would be taking less of a risk in becoming heavily dependent on China because other options would continue to be readily available. Moreover, a slack market would strengthen the Japanese hand in price negotiations. China is unlikely to join OPEC because it would prefer to retain its freedom of action and in any case (as explained in chapter 2) does not intend to export on a large scale. At the same time, Peking has frequently expressed its solidarity with OPEC as a champion of Third World interests and would prefer to avoid undercutting the cartel, including its only Asian member, Indonesia. China dropped the price for Taching crude to $12.10 under Japanese pressure in January 1975, thus undercutting the Indonesian price of $12.60, and then raised the price again to $12.30 after the OPEC price increase in November. In early 1977, Peking went up to $13.15, still forty cents below Indonesia. Nevertheless, the

Chinese price was substantially higher than the price of the Arabian "light" on which Japan primarily relies. Given the Chinese advantage in freight costs resulting from proximity, as the *Japan Petroleum Weekly* noted, Taching crude is inherently more profitable for Japan than Indonesian Minas crude.[19]

Another critical variable is likely to be whether China is able to export varieties of oil with a lower wax content than Taching crude from other fields located, like Taching, in coastal areas within feasible pipeline reach of ports facing Japan. Shuichi Kumano, president of Mitsui Oil Development, has said that "we have reason to think there is a fair amount of desirable light oil in southern China, both onshore and offshore." [20] It is not yet certain whether Takang oil contains significantly less wax than Taching. However, Masao Sakisaka, president of the Institute of Energy Economics in Tokyo, told me that some of the Takang oil tested in Japan is of "a light type desirable for us, and if Taching oil is replaced with this oil our capacity to absorb Chinese oil would be enlarged." Preliminary indications suggest that the grades of oil found so far in offshore areas of the Po Hai Gulf and the Yellow Sea are lighter than Taching crude. This is one of the principal factors explaining the intensity of Japanese interest in Chinese offshore progress.

Even if one assumes that China will reach anticipated production levels and will prove able to export a lighter grade of oil than Taching crude on a large scale, an important factor affecting export levels to Japan will be how soon Peking is able to expand its pipeline, tanker, and port capabilities. So far, large-scale harbor dredging and expansion efforts are under way at nine major ports. At Tientsin's port of Hsinkiang, wharves are under construction to accommodate 35,000-ton tankers. At the twin ports of Lushun and Talien (Dairen), known by the combined name of Luta, new loading facilities will accommodate tankers ranging from 50,000 to 100,000 tons in capacity. At Chin Huang Tao on the northwest side of the Po Hai Gulf, the terminus of a new pipeline from Taching, petroleum-loading berths and storage facilities are near completion. At the main southern oil port of Chanchiang, a 453-foot deep-water berth for 50,000-ton tankers has been completed. Other harbor expansion programs partially related to oil exports are in progress at Yentai, Chingtao, Lienyungkang, Shanghai, and Huangpu. Despite this massive expansion program, however, still more ambitious efforts would be needed to attain export levels to Japan surpassing the 18 million tons (135 million barrels) per year projected in the 1977 negotiations. More important, new pipelines would be required as new onshore and offshore fields go into production.

As for political variables, one significant imponderable is likely to be the progress of long-pending efforts to evolve cooperative energy development arrangements with the Soviet Union. While moving toward a much greater energy reliance on China than on the Soviet Union, as suggested earlier, Japan has not given up its hopes with respect to the Sakhalin and Yakutsk ventures and could well become uneasy if its position vis-à-vis the two Communist rivals should become too one-sided. This would be especially so in the event of a deterioration in Sino-Soviet relations posing the threat of war or, alternatively, a return to the close collaboration of two decades ago. Masao Sakisaka stressed the potential of the Yakutsk project, describing it explicitly as an offset to a total dependence on China that would give Japan greater scope for a continued escalation of oil imports from Peking.

By far the most critical political variable governing Japanese attitudes is likely to be whether China maintains its national unity in the post-Mao period and continues to impress Japan as a country on the way to eventual great-power status. Yasuhiro Nakasone, then minister of international trade and industry, stressed the "many-sided implications" of the oil link in a late 1974 interview, observing that "we have basic geopolitical reasons for wanting very substantial imports of Chinese oil, quite apart from our desire to diversify our energy supplies. A stable oil partnership will help to cement a close political relationship, and so we take a special interest in achieving such a connection." The Japanese image of

China as a future world leader was apparent in the upbeat comments of prominent business leaders who returned from a China visit in 1975 bubbling over with confidence that China had reached the takeoff stage in its development. In urging that Japan make its economy more complementary to that of China these leaders emphasized not only the economic advantages of doing so, but also their conviction that China would be an increasingly important factor in the world balance of power. Bunpei Otsuki, president of Mitsubishi Mining and Cement, pointing to China's riches in oil, iron ore, and coal, thought that it would become "an extraordinary country with great power," particularly if it enlisted the help of Japanese technology.[21] Toshio Doko, leader of the delegation and president of Keidanren, found the Chinese people "burning with hope" and predicted that Peking would become "a major power center in the world and in the Pacific sphere. An important problem for us will be deciding where Japan should be in the world context when that happens." Suggesting that China may become "a more promising customer for Japan than the United States" in the decades immediately ahead, Doko called for a restructuring of Japanese industrial plans to permit the import of Chinese industrial goods along with large-scale oil imports.[22] *Nihon Keizai*, echoing Doko's plea, stressed that China had become second only to the United States as a market for Japanese exports and linked the issue of oil imports directly with signs of economic trouble ahead in the West, declaring that "it is urgently necessary to put Japan-China trade on the road to stable and balanced expansion at a time when world trade is showing signs of stagnation and decline due to depression and inflation." [23]

Japan and the East China Sea: Ready and Waiting

Confronted with inherently uncertain economic and political factors in dealing with China, Japan is keeping its options open by cautiously developing a limited offshore exploration and production capability specifically designed for possible use on the East China Sea continental shelf. Japan is already a leading rig manufacturer, and Tokyo clearly has the know-how necessary to exploit the resources of the shelf on its own should it ever make a determined attempt to do so. Significantly, however, only a carefully circumscribed offshore program has so far been authorized by government and business leaders. Advocates of an accelerated Japanese offshore effort have until now been unable to obtain either the funds required for a major technological breakthrough in deep-water drilling or a political go-ahead permitting them to explore in sensitive parts of the shelf.

In a series of interviews with Japanese business leaders and officials, I found a widespread belief that so long as Sino-Japanese relations appear to be improving, it makes more sense to gamble on the continued flow of oil from China than to take the costly risks involved in offshore exploration. Morihisa Emori, vice-president of the Mitsubishi Research Institute, expressed the prevailing sentiment when he told me:

We could produce oil in the East China Sea if we wanted to because our people have had experience in offshore work in not only the Middle East but in such places as Malaysia, Sarawak, Burma, Vietnam, and Thailand. But the point is that it's not worth spending the money when we consider the opportunities we have elsewhere, especially in Southeast Asia, which do not involve comparable political risks. Taiwan is the heart of the problem. Until it is settled, we must wait and see.

Masao Sakisaka also emphasized that the situation is "suspended" until the Taiwan issue is settled and that

There are very big risks involved, which we wouldn't want to take unless we have no other alternatives. . . . While we have or could develop the technology necessary for exploiting the continental shelf, we do not have the experience, and it would not be easy. If we should seek to do so, the task would be too big for individual Japanese companies, and we would have to combine into a single national effort to distribute the risk.

But Sakisaka thought that China would export as much of its oil as possible to earn foreign exchange, restricting domestic consumption, "and there is no reason to assume, as some do, that they will limit their exports to an arbitrary level such as 10 percent of total production." Within "five to ten years," he said, it was conceivable that some form of cooperative venture might evolve with China on the shelf, although it would be preferable to have a median-line solution giving Japan a free hand on its side of the shelf.

Saburo Okita, director of the Overseas Cooperation Fund, stated:

We used to feel more strongly about controlling our sources of oil than we do now after OPEC. Now getting oil seems more important to us than developing it ourselves. Offshore work is very speculative and expensive. The main thing is getting it nearby and diversifying our sources. So while we would like to develop part of the shelf oil ourselves, under some sort of median-line solution, the main thing is the supply of oil. It's all right with us if China develops it, perhaps with American help and Japanese help, and we don't mind if U.S. companies find some oil for China and then sell it to us.

Underlining the costs of offshore development, Nobuyuki Nakahara, president of Toa Nenryo, affiliated with Mobil and Caltex, stressed that "only very big structures pay off. You need at least 100,000 barrels [1,330 tons] a day from an offshore structure to make it economic, and how often does that happen?" Akinobu Tsumura, research director of the Japan Petroleum Development Corporation, was one of several Japanese experts who said that natural gas appeared more likely than oil on the basis of geological studies on the shelf, which would make offshore production particularly expensive.[24] Yoshio Tanabe, chief China analyst in the Cabinet Research Office, the Japanese equivalent of the CIA, said that it would be easier to import oil directly from China than to invest money in the Senkakus (Tiao-yü T'ai) or other parts of the shelf, "so we would rather not excite the situation at this time. Let us just wait and see how things develop." Similarly, Kenichiro Maiya, a leading Japanese writer on oil affairs, pointed frankly to the flow of oil from China as the key variable, suggesting that Japan would calibrate its costs and benefits with cold-blooded precision:

It all depends on how much oil Japan gets from China. If it is 40 million tons [300 million barrels] per year, let us say, that is one thing, and the East China Sea will not seem very important. But if it seems that 20 million tons [150 million barrels] will be the maximum we can ever expect, then the Senkakus and the shelf will loom larger.

Supporters of a more assertive Japanese posture on the shelf openly warn that China, for its part, is consciously using oil exports to cool Japanese interest in independent offshore activity. "The fact that China has recently been active in its oil diplomacy toward Japan may not be totally unrelated to its interest in developing the East China Sea on its own," speculated *Zaikai Tembo,* a leading business review. China wants to buy equipment from Japan for its own offshore efforts, utilizing the foreign exchange earned from its oil exports, the journal declared, but will continue to rebuff Japanese efforts to arrange joint exploitation of the shelf. Citing Chinese support of proposals for an offshore economic zone in Law of the Sea negotiations, *Zaikai Tembo* warned:

China intends to expand its sphere of control over its seabed oil fields gradually after it has established its territorial waters in the East China Sea by posing as a supporter of the developing countries at the Law of the Sea Conference. Japan should not overlook China's true intentions hidden behind its increasing oil supplies to Japan.[25]

In weighing the possible risks and costs of exploring the East China Sea, Japan has been strongly influenced by the notably unsuccessful experience of Japanese companies to date in searching for offshore oil. This has been true overseas, where Japanese companies have suffered heavy losses in offshore joint ventures with foreign partners, especially in Indonesia. More important, the speculative character of offshore exploration has also been dramatically underlined at home by the spotty results of initial Japanese efforts to find oil and gas in close-in continental shelf areas immediately adjacent to the Japanese main islands. It was not until 1967 that Japan began to organize these domestic efforts under governmental direction and not until 1971 that the first serious drilling was undertaken in the Sea of Japan. In the five subsequent years, however, several Japanese oil companies and their foreign partners, backed to a limited extent by the Japanese government, spent $105 million and drilled forty-one holes, spurred by geological surveys pointing to probable recoverable reserves of 826 million tons (6.2 billion barrels). By the end of 1975 this exploration effort had resulted in the development of only one relatively small, commercially exploitable field with an estimated recoverable petroleum potential of 7.5 million tons (56.3 million barrels), or a daily production of 1350 barrels of oil and 1.6 million cubic meters of natural gas. The first discovery in the Aga field, located ten miles off the coast of Akita Prefecture in the Sea of Japan, was made during January 1973 in water 210 feet deep, and production there began in August 1976. One other problematical discovery was also made twenty-five miles off the Honshu coast near Joban in February 1973. While the Joban field was hailed as commercial by Teikoku Oil, the Japanese company involved, Teikoku's American partner, Exxon, has concluded that the water depth of 500 feet would make production unprofitable.

Notwithstanding this discouraging start, Japanese advocates of a more ambitious domestic exploration program have insistently argued that these early efforts have been too modest and haphazard to be conclusive and have pressed for all-out government subsidies to underwrite the risks taken by private companies. The "offshore lobby" in Japan is spearheaded by the business enterprises concerned, by geologists, and by the Development Department of MITI's Resources and Energy Agency. Long ignored, offshore enthusiasts have gradually begun to make an impact on official policy since the appearance of OPEC. In 1973, MITI's first energy white paper declared that "in order to guarantee Japan's oil supply it is necessary to encourage oil development overseas and on the continental shelf around Japan. Exploration and development of oil resources through her own efforts is Japan's most effective means . . . of guaranteeing an independent, secure supply source." [26] In 1975 the high-level National Energy Countermeasures Conference mapped efforts to reduce Japanese dependence on imported oil to 63 percent of the overall national energy supply by 1985. This would be a sharp drop from a 77.4 percent dependence in 1973 and envisaged a stepped-up offshore program in addition to greater efforts in nuclear and coal development.

The Ministry of International Trade and Industry took the first significant steps to set the stage for greater offshore efforts by winning Diet approval in 1975 of a bill empowering the Japan Petroleum Development Corporation to make direct government loans to companies exploring for offshore oil in domestic waters. The bill was accompanied by a $117-million allocation for such loans to the Japan Development Bank in the 1975 fiscal year alone. MITI also won approval in the 1975 budget for expenditures in other areas of offshore research and development totaling $145 million. In the 1976 budget, offshore development expenditures claimed an additional $48.9 million, including a $24.3-million allocation to the Japan Petroleum Development Corporation for seismic surveys and other activities under its direct authority. [27] In late 1976, eleven oil companies were at work exploring for offshore oil near Japan.

To some extent, these expenditures are intended to maximize Japanese options for independent development in disputed portions of the East China Sea shelf should future political circumstances permit. Significantly, however, MITI also has other purposes in mind. For one thing, Japanese drilling companies are likely to be increasingly active on a global scale, and domestic exploration offers badly needed practical experience. The report of a 1975 conference of company and government officials interested in offshore activity stressed this theme, declaring:

Since Japanese companies are themselves taking the responsibility for oil development on the continental shelf, in contrast to our previous work in Japan and overseas with primary reliance on foreign partners, we will accumulate our own know-how and experience in this important field, relying less and less on foreign technical help, which will contribute greatly to Japan's independent role within the global arena in addition to increasing our oil supplies.[28]

In 1975 Japan had only 101 technicians with offshore drilling experience. Most Japanese oil companies were refiners and sellers (operating "downstream") rather than explorers (operating "upstream") as a result of the postwar Japanese dependence on the majors. "We were spoiled," reflected Foreign Minister Kiichi Miyazawa, who had oil exploration efforts under his jurisdiction as minister of international trade and industry from 1968 to 1970. "We thought we had oil for the asking during all those years," he explained in an interview, "and we were very slow in realizing the importance of training our own people and searching for our own reliable sources of supply."

By channeling loans into domestic exploration, the new MITI program upgrades the operational capabilities of Japanese companies, and by sponsoring futuristic offshore research it also seeks to keep them abreast of the latest Western technology. A semisubmersible drilling rig now under construction, the *Hakuryu* ("White Dragon") *V*, will be capable of drilling in water 1,650 feet deep and will have "dynamic positioning." A little-noticed but significant item in the 1976 budget also provided $521,000 for the design of a "super-rig" capable of drilling 3,000 to 6,000 feet deep, equaling or outdoing such avant-garde U.S. rigs as the *Glomar Explorer*, the *Challenger*, and the *Sedco 445*. Under licensing and subcontracting arrangements with American partners, Japanese shipyards have been making increasingly sophisticated rigs, among them the self-propelled floater *Discoverer III* for the Offshore Company, capable of operating at a depth of 1,000 feet, and the *Ocean Ranger*, a semisubmersible built for the Ocean Drilling and Development Company with a unique chain-and-wire mooring system designed to permit drilling in water depths of up to 3,000 feet. As of June 1975, a total of twenty-five drillships, semisubmersibles, and jack-up rigs had been delivered by Japanese shipyards, and construction was started on another twenty-three during late 1975 when supertanker orders dropped off.

In the past, Japan has been dependent on the United States for key components of its rigs, especially the generators and drawworks, but the MITI program is seeking to reduce this dependence. Similarly, MITI sees domestic offshore efforts as a stimulus to acquiring experience in manufacturing and using production platforms. The platform installed by Nippon Steel at Aga in 1974 was the first full-scale platform ever put into operation by a Japanese company.

In developing its rig technology, Japan has in mind worldwide markets and contracts as well as its domestic offshore activity. In addition to its rig program, however, Tokyo has also been upgrading its seismic survey capabilities and has taken several limited but significant steps that appear to be specifically related to the possible future development of the East China Sea shelf. In 1974 the *Takuyu Maru* ("Ocean Explorer"), a computer-equipped vessel developed and operated by the semigovernmental Japan Petroleum Development Corporation, made an unannounced seismic survey reaching more than one hundred miles onto the shelf (Figure 6). In August and September of 1975, *Hakurei Maru* ("White Ridge"), an oceanographic research vessel operated by the Geological Survey of Japan, was assigned to prepare a submarine geological map covering the entire shelf. Tokyo had hoped that the project would prove politically inoffensive, since it did not entail advanced seismic surveys, but the ship hastily turned back after Taiwan got wind of the venture and protested. A more cautious survey in waters immediately west of Okinawa was then conducted by the *Takuyu Maru* in late 1975. Covering 3,100 miles of seismic "lines," the survey embraced areas on the continental *slope* where the continental shelf drops downward into the Okinawa Trough, the oceanic canyon dividing the shelf from the Ryukyu Islands (Figure 6).

According to one school of thought in international law, a continental shelf includes not only its slope but also the *rise* at the bottom of the slope, where the accumulations of potentially oil-bearing sediment are believed to be even greater than on the shelf proper. As part of any claim to the shelf, China could conceivably lay claim to the slope and rise in the Okinawa Trough. The 1975 seismic survey west of Okinawa was part of a larger, three-year program to survey 15,000 miles of seismic lines on the slope that included areas near Miyakojima and other islands at the southwestern tip of the Ryukyuan chain in addition to areas near the Japanese main islands. This program is to be stepped up following completion of Japan's biggest and most advanced seismic survey ship, *Kaiyo Maru* ("High Seas"). The report of the 1975 offshore planning conference, cited earlier, underlined the potential of the slope, observing that "when drilling and development companies move onto the slope in the future, the target area will be far wider than the continental shelf." [29] Moreover, the futuristic super-rig with a 3,000- to 6,000-foot drilling capability now under development by the Japan Petroleum Development Corporation is being designed explicitly for use "in areas of the shelf and the slope judged to contain oil and gas reserves." [30]

The Senkaku Islands
in Suspension

For the present, except in the areas surveyed near Okinawa, Japan does not appear to be planning further seismic surveys in shelf areas where China might eventually make boundary claims. Prior to the opening of U.S. and Japanese contacts with Peking, however, Tokyo had embarked on a serious program for the development of the shelf, even allocating provisional concessions to Japanese companies in 1969 and 1970 on the eastern side of a hypothetical median line based on the Japanese claim to the Senkaku Islands (Tiao-yü T'ai). Since the Japanese suspension of this program in 1972, private Japanese companies have not been permitted to pursue seismic surveys or exploratory drilling in contested areas, and as noted above, government-sponsored vessels have acted with great caution. Nevertheless, it should be kept in mind that Japan retains serious and long-standing claims to a share in the riches of the shelf. Even before World War II, Japanese geologists had been fascinated by the oil potential of the East China Sea,[31] and this interest was heightened by the wartime findings of Japanese and U.S. oceanographic studies conducted in connection with submarine activity. Kazuo Yatsugi, a prominent figure in the Tojo regime, recalled in an interview that Japanese oil experts had conducted exploratory shallow-water drilling in the Senkaku (Tiao-yü T'ai) area during the war and had found evidence of natural gas. As mentioned in chapter 3, Professor Hiroshi Niino of Tokai University, in collaboration with K. O. Emery of the Woods Hole Oceanographic Institution, began as early as 1961 to expand upon the data compiled during the war. By 1967 Niino and Emery had published their study describing the East China Sea as "one of the most potentially favorable but little investigated" of the world's continental shelves. Largely unnoticed outside Japan and South Korea, this study aroused considerable behind-the-scenes excitement in Japanese governmental circles and added to the intensity of the Japanese pressures then building up for the return of Okinawa by the United States.[32] What stirred special interest in the context of the reversion issue was a finding emphasizing the potential of undersea areas directly flanked by Okinawa. "The most favorable province for future submarine oil and gas fields," Niino and Emery wrote, "is a wide belt along the outer part of the continental shelf." [33]

Japanese interest was focused particularly on the southern part of the East China Sea, where an Okinawan entrepreneur, Tsunenobu Omija, had been sponsoring his own geological studies in the area near the Senkakus (Tiao-yü T'ai) for years and had long claimed to have found evidence of major oil deposits. In July 1968, when Prime Minister Sato was just beginning to push the Okinawa issue, Daisuke Takaoka, a prominent Liberal Democratic party political figure and a member of an advisory group on Okinawan affairs in Sato's office, led a small geological expedition to the Senkakus (Tiao-yü T'ai).[34] Several months later, Niino and Emery, following up the 1967 study, conducted their much-publicized United Nations seismic survey of the East China and Yellow seas, prompting more comprehensive Japanese studies focused on the area surrounding the Senkakus (Tiao-yü T'ai) in 1969 and 1970 under the direct sponsorship of the cabinet. The first of these, covering a 24,840-square-mile area, proved inconclusive.[35] In 1970, however, the results of intensive studies in a somewhat smaller area were more encouraging, and plans were drawn up for a five-year crash program to search for oil in the southern part of the shelf.

The high hopes lying behind this projected exploration effort were apparent in a revealing roundtable discussion held on 20 November 1970. Yutaka Ikebe, director of the Japan Petroleum Development Corporation, hailed the potential of the East China Sea as "comparable to Saudi Arabia," and geologist Michihei Hoshino of a Tokai University survey team, then just back from the Senkakus (Tiao-yü T'ai), predicted that the East China Sea shelf would rank as "one of the five biggest oil producing regions of the world." Ikebe declared that further seismic studies would be conducted in the area during 1971 and would be followed by the first exploratory drilling in 1972 or, "at the latest, by 1974 or 1975." During this time span, he said, Japan would have little difficulty in upgrading its drilling capabilities to the extent required, but special efforts would have to be made to master the problems associated with production in the Senkaku (Tiao-yü T'ai) area, where the water depth averages 390 feet and reaches 510 feet in some places. When Tatsunao Takaha-shi of MITI outlined a five-year program for surveying the entire East China Sea shelf, Hoshino urged companion surveys of the continental slope probing "deep down between the continental shelf of Okinawa and the East China Sea." Ikebe assured Takahashi that the government had plans for the slope, among them the development of a $5-million seismic survey vessel equipped to work in deeper waters than existing Japanese vessels. It was this ship, *Takuyu Maru*, that began the cautious probes of the shelf near Okinawa in 1975 already noted. Asked during the discussion how long it would take to catch up with U.S. technology in the offshore field, Takahashi said that Japan was ten years behind the United States, but Hoshino felt that Japan could develop technology adequate for the Senkakus (Tiao-yü T'ai) and surrounding areas by 1975. Ikebe pointed out that the *Hakuryu III*, then under construction, would be finished by May 1971 and would soon be at work in the Sea of Japan. "It will keep on practicing there until 1975," Ikebe added, "and then it will be easy to take it to the Senkaku area for full-fledged operations." [36]

Japan's aborted plans for development of the East China Sea were promoted not only by MITI officials and those directly concerned in the oil industry but also by rightist, pro-Taiwan elements of the Liberal Democratic party who saw offshore oil development as a vehicle for an anti-Communist regional alignment under Japanese auspices. In July 1970, well before Taiwan had signed its concession agreements with Western companies, a group of rightist Japanese business and political leaders, led by former prime minister Nobusuke Kishi, unveiled a plan for a tripartite oil-development program to be jointly sponsored by Japan, Taiwan, and South Korea. This program was to be the centerpiece of a larger effort to expand the activities of the long-standing Japan–Republic of China Cooperation Committee into those of a Japan–Republic of China–Republic of Korea Liaison Committee. Disturbed by ever-louder rumblings of a change in U.S. China policy, Kishi and his allies felt that it was urgently necessary for like-minded elements in Taipei, Seoul, and Tokyo to create a solid economic partnership. In October a MITI vice-minister and several Japanese oilmen held unannounced discussions with Taiwan officials concerning two different proposals for joint continental shelf development. One of the proposals envisaged joint operations in a 6,210-square-mile area of the Taiwan Strait in Taiwan Zone I, one of the few offshore Taiwan areas that had not been covered in the September agreements with U.S. companies.[37] The other would have involved a cooperative approach in the area surrounding the Senkakus (Tiao-yü T'ai) already granted to Gulf under the September agreements.

Although the precise terms of this plan were never revealed, it is noteworthy that the Japanese participant would have been Teikoku Oil.[38] Teikoku and Gulf have a relationship in Japan that was then and continues to be one of considerable importance to Gulf, involving offshore areas far more extensive than those held by Gulf in Taiwan Zone II. The plan apparently called for deferring efforts to settle the Senkaku (Tiao-yü T'ai) dispute between Tokyo and Taipei and dividing any spoils of oil development through arrangements assuring the harmonization of the private corporate interests involved. Thus, Teikoku and Gulf would have had a Japanese concession overlapping the previously existing Gulf concession from Taiwan, which would have been renegotiated to restructure Gulf's obligations to Taipei.

This and similar ideas never got off the ground because Peking reacted sharply when Kishi's tripartite liaison committee publicly announced its plans for offshore development of the shelf at meetings in Seoul and Tokyo in November and December. Taking note of the committee for the first time, the *Peking Review* branded its creation "a new crime committed by Japanese militarism in plotting aggression against China with U.S. imperialist support, a serious provocation by the U.S. and Japanese reactionaries against the Chinese and Korean people."[39] Three months later China insisted on explicit repudiation of the committee by Japanese delegates to the annual Sino-Japanese trade talks in Peking. In the communiqué accompanying the Memorandum Trade Agreement signed on 1 March 1971, China declared:

The newly-established Japan–Chiang–Pak "Liaison Committee" has gone so far as to decide on the "joint exploitation" of the resources of the shallow seas adjacent to China's coasts. This is a flagrant encroachment on China's sovereignty, and the Chinese people absolutely will not tolerate this. . . . The Japanese side states that it understands this solemn stand of the Chinese side [and] pledges to struggle resolutely against these reactionary activities encroaching on China's sovereignty.

In canceling its plans to conduct further surveys of the East China Sea shelf during 1971, Japan was motivated by the vehemence of the Chinese stand and, above all, by the turnabout in U.S. China policy initiated in that year. The shifting U.S. posture was a cataclysmic development in Japanese eyes, calling for a comparable volte-face on the part of Japan. Peking had formally reasserted its own claim to the Senkakus (Tiao-yü T'ai) in response to the claims by Tokyo and Taipei, and Tokyo quickly reacted by offering to pigeonhole the issue as a means of advancing the normalization of Sino-Japanese relations. At the same time, Japanese leaders were not prepared to see Taiwan move into the resulting vacuum by means of its agreements with American companies. Once plans for joint development with Taipei had been abandoned, Tokyo moved decisively to frustrate exploration by Gulf in the Senkaku (Tiao-yü T'ai) area under its agreement with Taipei. Stymied by the official U.S. disapproval of survey activities by the U.S. vessel *Gulfrex* related in chapter 1, Gulf had engaged a West German survey boat, the Prakla Company's *Korax*. When the *Korax* began to explore in the Senkaku (Tiao-yü T'ai) area, the boat carefully stayed outside a twelve-mile arc around Uotsurishima, the largest of the Senkaku (Tiao-yü T'ai) islands, hoping to avoid provocation to Japan by showing respect for the Japanese territorial claim to the islands as such, while at the same time asserting Gulf's exploration rights under its concession agreement. Japan responded quickly, however, dispatching an armed Maritime Agency reconnaissance vessel, the *Satsuma*, on 7 June 1972 to photograph the boat's operations. Tokyo, anxious to avoid a public fuss in view of the backstage preparations then getting under way for the midsummer visit of Prime Minister Tanaka to Peking, handled the affair through quiet diplomatic pressures on West Germany and the United States. Gulf finally withdrew the boat on 11 June 1972 to a less controversial portion of its concession farther west.

Spurred by Japan's strong reaction, Gulf eventually relinquished the portions of its Taiwan concession near the Senkakus (Tiao-yü T'ai) under a contractual timetable providing for the periodic selection and relinquishment of designated percentages of the original concession area. Taiwan has since kept this an "open" zone, withheld from development either by foreign concessionaires or by its own government oil company, the Chinese Petroleum Corporation. Japan, for its part, has winked at seismic surveys in Taiwan concessions overlapping provisional Japanese concessions except in this instance involving the area near the Senkakus (Tiao-yü T'ai). Tokyo apparently sees no damage to its interests in letting seismic surveys proceed, but there are strong indications that it would take a sterner view of exploratory drilling. When Superior was on the verge of drilling in Taiwan Zone IV in February 1975 (as discussed in chapter 5), Teikoku Oil sent an implicitly threatening letter asserting its claim to the area. So far, no drilling has taken place in Taiwan in areas overlapping the provisional Japanese concessions. Indeed, no drilling at all has occurred in concessions north of Taiwan with the exception of drilling by Gulf relatively close to the island.

The fact that Gulf has carefully confined its drilling to the southernmost portions of its Taiwan concession merits special emphasis. Some company officials state that this has been primarily for geological reasons and because production is less expensive close to shore. However, political considerations may well have played a part in the selection of the three sites where drilling took place during 1974 and 1975, since the provocation to China and Japan alike clearly increases in direct proportion to the distance from Taiwan. Significantly, Gulf appears even more concerned about offending Japan than China, given its extensive stake in the Japanese offshore concessions already noted. In partnership with Teikoku, Gulf has an interest in seventy-three million acres of offshore territory in its Japanese concessions, in contrast to a total of only nine million offshore acres in the United States. Most of this acreage is located in provisional concessions on the shelf, but some of it is in areas close to the Japanese main islands, notably Amakusa and Miyazaki, and some is in areas of the shelf and slope near Okinawa. In April 1975 Gulf and Teikoku formed a special exploration arm, Teiseki Tairiku-dana Kaihatsu (Teiseki Continental Shelf Development Company), launching seismic surveys during 1975 in unspecified areas of "the seas around Okinawa." [40]

The area where Japan is most eager to resume its interrupted program of offshore development is substantially the same as the 24,840-square-mile zone surrounding the Senkakus (Tiao-yü T'ai) that was originally embraced in its 1969 and 1970 surveys. The western part of this area has been provisionally allocated to the governmental Japan Petroleum Development Corporation, and it is this concession that overlaps partly with Gulf's Taiwan Zone II. To the east of this concession, another promising area has been allocated to Uruma Resources, backed by the Nissho-Iwai trading combine in alliance with Tsunenobu Omija, the Okinawan businessman who claims to represent Okinawan interests as against those of the Japanese main islands. [41] The Uruma concession overlaps with the suspended, open Taiwan zone immediately adjacent to the Senkakus (Tiao-yü T'ai) and with Taiwan Zone III to the north, so far allocated by Taipei to Oceanic Exploration (Figure 6).

Japan's claims to these and other areas on the East China Sea shelf hinge upon its title to the Senkakus (Tiao-yü

T'ai). Given the location of these islands on the shelf, recognition of Japanese sovereignty over the Senkakus (Tiao-yü T'ai) would establish Japan as a shelf power entitled to seek a median line fixed in relation to the Senkakus (Tiao-yü T'ai) as its easternmost base point. In the absence of such recognition, China could press its claim to the entire shelf as a natural extension of Chinese territory and refuse to accept the principle of a median line.[42] Despite this, as the concluding chapter will elaborate, Peking has left the door open for cooperative arrangements with Japan that could eventually permit Japanese-oriented oil development to go forward in the southern part of the shelf pending a formal resolution of the Senkaku (Tiao-yü T'ai) dispute. Some form of compromise arrangement is not inconceivable in the context of increasingly close Sino-Japanese relations, just as a conflict over the Senkakus (Tiao-yü T'ai) cannot be ruled out should relations between Tokyo and Peking deteriorate.

As for the short-term future, a key official of the Japan Petroleum Development Corporation could only complain bitterly in 1975 that "every other country so far interested in doing so has conducted its own exploration of the East China Sea continental shelf" to follow up the promising findings by the United Nations in its 1968 survey and that thus far "only Japan has been unable to do so." Summing up the prospects for developing the Senkakus (Tiao-yü T'ai), Uruma Resources told the 1975 Tokyo offshore planning conference that "if and when permission is obtained from the proper agency, geological testing is planned." [43]

Oil, the Right Wing,
and the China Connection

The sensitivity of the Senkaku (Tiao-yü T'ai) dispute as a possible flash point in Sino-Japanese relations is largely a byproduct of the struggle over the resources of the continental shelf. In manipulating the issue, however, both sides have sought to use it as a reminder of past triumphs or humiliations. Japan traces its claim to the islands back to their capture following the 1894 defeat of the Ch'ing regime in the first Sino-Japanese war. It is the Japanese contention that the Shimonoseki peace treaty of 1895 implicitly treated the Senkakus (Tiao-yü T'ai) as Japanese territory by failing to include them as part of Taiwan when Taiwan was ceded to Japan under the accord.[44] Subsequently, Japan argues, the United States confirmed the Japanese claim by administering the Senkakus (Tiao-yü T'ai) as part of the Ryukyus after World War II and by returning them as part of the Okinawa reversion agreement.[45] China challenges the Japanese position by asserting that the Senkakus (Tiao-yü T'ai) had been part of the Chinese tributary system in East Asia for some five centuries before the Japanese took control. Although never inhabited, China maintains, they were discovered by China and were long considered a part of Taiwan.[46] For the Japanese Right, the Senkakus (Tiao-yü T'ai) provide a vestigial symbol of erstwhile imperial glory, while for China they evoke memories of the setbacks suffered when the nation was weak and divided.

The emotional appeal of the Senkaku (Tiao-yü T'ai) issue has been magnified by the fact that there are no other territorial issues that can serve as a focus for the feelings of hostility between China and Japan left over from the 1894 conflict and the Japanese occupation of northern China from 1937 to 1945. This appeal was strikingly apparent in the breadth and intensity of the anti-Japanese protests over the Senkaku (Tiao-yü T'ai) controversy organized by Chinese students in Taiwan, Hong Kong, and the United States during the 1970–72 period. In the case of Japan, such intense emotions are notably absent, with strong feelings regarding the Senkakus (Tiao-yü T'ai) limited to right-wing elements of the Liberal Democratic party. Nevertheless, the issue continues to be significant in Japanese politics because the Right has attempted to use it as a means of complicating negotiations for a Sino-Japanese peace treaty and thus of slowing down the overall progress of relations between Tokyo and Peking.

In assessing the potential of the Senkaku (Tiao-yü T'ai) issue, it should be remembered that China does not occupy the islands and that Japan has the closest military access to the area from the nearby Ryukyu Islands, occasionally sending naval patrols into the area. A peace treaty omitting mention of the Senkakus (Tiao-yü T'ai) would not weaken the Japanese posture in relation to the islands and might even be viewed as a partial Chinese retreat. By contrast, the Soviet Union does have military control of four disputed islands to the north of Japan, and a peace treaty that failed to confront the "northern islands" issue would constitute, in effect, an abandonment of Japanese claims.[47] So long as Peking does not attempt to annex the Senkakus (Tiao-yü T'ai), it appears unlikely that Japan will press its territorial claims, although this posture would not rule out Japanese efforts to get Chinese approval for some arrangement permitting oil development to proceed.

Significantly, even the Japanese right wing has lost some of its interest in the Senkakus (Tiao-yü T'ai) as the Sino-Japanese relationship has grown more intimate and has begun to produce a payoff in the form of Chinese oil exports. Kazuo Yatsugi, a long-standing prime mover in the Taiwan lobby, queried:

If China can supply us with oil, why should we stir up troubled waters? It is good to have sources of oil nearby, but we shouldn't risk burning our fingers in the political fires. It is more sensible to be patient and wait until political complications are settled. We can't turn the clock back regarding Taiwan, and we can't settle the territorial issue without negotiations.

In light of the Sino-Soviet rivalry, Yatsugi said in an interview, "we have had to enter into close relations with China in order to help China to keep a balance with the Soviets."

In contrast to the enthusiasm for joint oil development

efforts with Taiwan shown during 1970, Japanese leaders have consistently rebuffed intermittent overtures for such arrangements since Tokyo began to alter its stance toward Peking in mid-1971. To U.S. firms with concessions granted by Taipei and to Japanese firms holding overlapping provisional concessions from Tokyo, overt or covert arrangements opening the way for oil exploration seem eminently logical. Clinton International, for example, has made repeated approaches to Japan since acquiring its concession in late 1970, principally through former Treasury secretary Robert B. Anderson, whose central role in Clinton's relationship with Taipei has been described earlier. As related in chapter 5, Anderson came to Tokyo in October 1971 seeking a Japanese concession for Clinton that would have coincided with Clinton's Taiwan concession in much the same way that the Gulf and Teikoku concessions would have coincided in the plan cited earlier. By early December the Japanese ambassador in Washington, Nobuhiko Ushiba, had firmly rejected the proposal in a letter to Anderson that provided one of the earliest signs of the sea change then taking place in the Japanese posture toward Peking. In the absence of any international agreement "between the parties concerned" on the demarcation of East China Sea boundaries, wrote Ushiba, "it is not appropriate for the Government of Japan to be engaged in such an arrangement with the Clinton International Corporation as you proposed." Japan does not accept Taipei's claims in the East China Sea, Ushiba added, but "even when it is established as a result of talks between the parties concerned" that Clinton's concession area belongs to Japan, there would be "little likelihood" of a Japanese concession for Clinton because "a Japanese company has already submitted an application for mining rights in the same area." [48] In a meeting with U.S. Ambassador Walter P. McConaughy in Taipei (cited in chapter 5), Clinton's managing director, Ted C. Findeiss, urged that the United States put pressure on Japan to have a Japanese company enter into joint arrangements "whereby essentially we would each cross-assign an undivided one-half interest in

each other's license." From Clinton's point of view, he reasoned, "half a loaf would be better than none," and in terms
of "free world interests, it is so logical for the governments
concerned to get together that there should be no question
about it." [49]

So far as can be determined, Washington has made no
effort to facilitate such joint exploration efforts in the years
since this proposal. However, both U.S. and Japanese companies with a stake in East China Sea concessions, Clinton
included, have continued to seek a basis for Tokyo–Taipei
collaboration. Among the examples of which I have evidence,
Anderson attempted to persuade Kyushu Oil Development to
seek an overlapping Japanese concession and to bring pressure on the Japanese government to change its attitude. As
late as 1973, Anderson was still pushing his cause in Tokyo,
as noted in chapter 5. Oceanic Exploration was also actively
seeking Japanese collaboration as late as 1974, approaching
Idemitsu Kosan, Kyushu, and other firms. At one point,
Oceanic was asking $22 million for a 12 percent interest in its
concession, "a price that was much too high," as an official of
one of the Japanese firms put it, "considering the political
risks involved." Presumably, any such arrangements would
remain unannounced and could be handled indirectly
through dummy companies abroad. An Uruma Resources
official told me that he had been approached by a leading
executive of Taiwan's Chinese Petroleum Corporation in
mid-1974 with a new variant of previously discussed plans
for covert collaboration.

In 1970 political support in Japan for the idea of a joint
Tokyo-Taipei approach to oil development had come from
rightist leaders who saw Taiwan as their major target of overseas interest. With the pro-Peking shift in Japanese policy,
however, the Taiwan lobby has progressively declined in
vigor, and there has been a pronounced shift of emphasis
from Taipei to Seoul on the part of rightist elements. Like
Taiwan, South Korea has not only been an attractive hunting
ground for Japanese investors and Japanese tourists but has

had a special political appeal. To the extent that Japan becomes entrenched once again in its former colonial terrain, rightist leaders of the older generation find a modest vindication of their shattered imperial dreams. Kazuo Yatsugi is one such rightist leader who has consistently promoted a little-known plan for the industrial integration of Japan and South Korea that closely resembles prewar blueprints for the "Co-Prosperity Sphere." [50] Joint oil development has increasingly become a central element in this plan since Yatsugi unveiled his proposal for tripartite oil development in November 1970. When cooperation with Taiwan was abandoned during 1971, Yatsugi, as convenor of the tripartite group, promptly began pushing the idea of Japanese–South Korean offshore oil cooperation. This ultimately resulted in the 1974 joint development treaty between Tokyo and Seoul discussed in chapter 6, a controversial agreement that was stalled for three years in the Japanese diet. On the one hand, the treaty has been attacked by China, North Korea, and Japanese critics of the Park regime; on the other, it is passionately supported by Japanese advocates of energy independence and by rightist elements who see it as a means of strengthening South Korea in its rivalry with the North. In the case of Taiwan, Japan had reacted quickly when concessions were given to U.S. firms in areas it claimed; similarly, one of the factors that spurred the Seoul–Tokyo agreement was a desire to assert control over American concessions in waters contested by Japan. In a 1971 memorandum, Yatsugi warned that "if Japan and South Korea permit other countries to develop their resources, this means that the interests of these outside countries are thrust into their territorial waters, which could cause troublesome problems." [51]

The political power of rightist elements in Japan continues to be considerable and has been strengthened by the successful 1977 effort to obtain Diet ratification of the offshore oil agreement with South Korea. Ratification has not only improved the morale of the Right; should the agreement with Seoul ever be implemented, notwithstanding Chinese and

North Korean objections, it could also bring concrete benefits in the form of important contracts for a variety of oil-related items ranging from steel pipe to helicopters. Much of the financial strength of the Right has come from a close working relationship with business leaders in which business dealings have been helped along by rightist political connections with the ruling groups in Taipei and now, to an even greater extent, in Seoul as well. For the Japanese firms involved, there have been profitable contracts, and for key political figures in Tokyo and Seoul, substantial rakeoffs. Kazuo Yatsugi and his mentor, former prime minister Nobusuke Kishi, have been closely allied in directing the Seoul lobby with Yoshio Kodama, the ultranationalist Tojo confidant who was convicted as a class-A war criminal during the U.S. occupation but reemerged to become a major wheeler-dealer in Liberal Democratic party affairs, as the Lockheed payoff scandal of 1976 revealed. Kodama has built his power base largely through his links with underworld gangs, some of them dominated by Koreans and closely tied in with influential groups in Seoul. The most important of these Korean gangs is the Tosei-kai, now underground, headed by Chong Kyu-young, known by the Japanese name of Hisayuki Machii, who presides over a multimillion-dollar conglomerate (TSK-CCC) embracing cabarets, bars, resorts, and massage parlors. In 1974 Tokuma Utsonomiya, chairman of the Audit Committee in the lower house of the Diet and a leader of the anti-Seoul, pro-Peking forces in Liberal Democratic ranks, elicited testimony showing that Machii owns the Pusan-Shimonoseki ferry which, according to authoritative sources, has been utilized to smuggle proscribed amphetamines from South Korea to Japan. Through legitimate and illicit channels, in short, rightist political forces in Japan and the Park regime in South Korea have forged tight interstitial ties, and this seems likely to ensure a continuing financial base for rightist activities.

Will the Right prove powerful enough to secure full-scale implementation of the Seoul–Tokyo treaty in the face of

Chinese opposition? All things considered, given the favorable prospects for a Sino–Japanese oil partnership, this appeared questionable in mid-1977. Some observers have predicted that China might acquiesce in implementation of the treaty as part of its anti-Soviet strategy, but there have been no signs yet of such an attitude. Moreover, it should be recalled that Japanese Foreign Ministry officials bent over backward to avoid provoking China when they negotiated the 1974 agreement with Seoul. As Figure 7 shows, the outer boundary of the joint development zone was fixed to the east of the boundary claimed by South Korea at the insistence of Japan. Seoul bowed to Japanese wishes on the understanding that the boundary of the joint zone as defined in the treaty did not represent a surrender of South Korea's independent claim to a previously demarcated boundary of its own farther west. When China protested the agreement anyway, demanding a voice in any delimitation of the shelf, pro-Peking forces in Japan launched their drive to prevent ratification. Should Peking insist on the natural-prolongation criterion, as in its 13 June 1977 statement, it could conceivably claim a portion of the zone bigger than that shown in Figure 7.

Gradually Japan has recognized that Peking is not only asserting its own interests in seeking to paralyze the 1974 agreement but is also championing the cause of Pyongyang as the only legitimate government on the Korean peninsula. By opposing the agreement, Peking appears to be serving notice that a joint approach to the continental shelf can only evolve as part of a larger political rearrangement in which North and South Korea are both cooperating parties to a new and broadened treaty. Similarly, as suggested in chapter 11, China is also likely to condition any Senkaku (Tiao-yü T'ai) agreement on Japanese support for a new Taiwan formula under which Taipei acknowledges the suzerainty of the mainland.

CHAPTER EIGHT

Danger Zones in the South China Sea

In the East China Sea the territorial dispute over the Senkaku Islands (Tiao-yü T'ai) is largely an extension of the emerging Sino-Japanese struggle for control of the oil and gas resources of the continental shelf. It was the competition for offshore resources that sensitized the dormant Senkaku (Tiao-yü T'ai) issue in the first place, prompting both Tokyo and Peking to utilize the islands as political symbols. By contrast, the South China Sea is the scene of a more complex power game in which conflicts of interest over resources, while potentially serious, are incidental to long-simmering territorial disputes that have now become closely interwoven with the Sino-Soviet rivalry for strategic advantage. Even if the South China Sea did not offer the long-term geological promise discussed in chapter 3, China would still have a significant political and military stake in the Paracel and Spratly islands. This chapter seeks to distinguish, accordingly, between the potential for conflict in the South China Sea resulting directly from Peking's offshore oil ambitions and the broader implications of the extensive sea boundary claims associated with the Chinese stand on the Paracels and Spratlys. Stretching all the way to Malaysia, Brunei, Indonesia, and the Philippines, these Chinese claims could not only hamper or even paralyze offshore development by the littoral states but could also have important political reverberations as well.

To the extent that there is a short-term danger of a collision over resources between Peking and its South China Sea neighbors, this danger would appear to be most serious with respect to Vietnam. Hanoi has offshore ambitions in the Tonkin Gulf that could lead to tensions with Peking in the relatively near future. In time, contending offshore claims could aggravate the continuing conflict over the Paracels and Spratlys, but it is equally possible that the two Communist allies will find a peaceful resolution of their interrelated resource and territorial disputes. It should be remembered that the Sino-Vietnamese conflict over the South China Sea islands was consciously rekindled by Saigon in 1973 as an offshoot of the Vietnam war. The Thieu regime attempted to bolster its sagging position by using a conflict with China to rally nationalist sentiment. In pressing Vietnamese claims to the Paracels and Spratlys, Thieu was also dramatizing hopes for an offshore bonanza that were increasingly emphasized during the last months of his regime to fortify popular morale. Yet ironically the oil discoveries of late 1974 and early 1975 may well have accelerated Hanoi's decision to launch its triumphant final assault against Saigon.

The Oil Factor
in Perspective

In the case of the Senkaku Islands (Tiao-yü T'ai), it is their location on the continental shelf that is significant to Peking and Tokyo. In the South China Sea, the continental shelf is relatively narrow, and the incipient conflicts between China and its neighbors are focused not on the shelf but rather on control of some 127 scattered and largely uninhabited coral atolls, cays, and sandspits (Figure 9). The most significant of these disputes concern the Paracel Islands (often identified on maps by their Chinese name, Hsi-sha, or their Vietnamese name, Hoang-sa), located 150 miles southeast of Hainan Island; the Spratly Islands (Nan-sha in Chinese and Truong-sa in Vietnamese), 550 miles farther south; and the Tseng-mu Reef, lying just 20 miles off the Malaysian state of Sarawak. The Paracels, claimed by both China and Vietnam, were the site of a much-publicized clash between Chinese and Vietnamese forces in January 1974 and have since been controlled by Peking. The Spratlys are claimed in whole or in part by China, Taiwan, Vietnam, and the Philippines; some of them have been occupied in recent years by token forces from Taiwan, Vietnam, and the Philippines, but the archipelago as a totality is not under the military control of any of the powers concerned. Three other significant areas claimed by China are the Pratas Reef (Tung-sha in Chinese), off Hong Kong; the Macclesfield Bank (Chung-sha), southeast of the Paracels; and the Scarborough Reef (Huang-yen), 160 miles southeast of the Macclesfield Bank. Chinese title to the Pratas Reef and the Macclesfield Bank does not appear to be contested, but the Scarborough Reef could become a potential focus of dispute with the nearby Philippines.

The sensitivity of the South China Sea as a strategic waterway is apparent from its location. To the southwest it connects with the Indian Ocean through the Malacca Strait, and to the northeast it commands access to the East China Sea. The sea lane running between the Paracels and Spratlys is used by oil tankers moving from the Persian Gulf to Japan as well as by warships en route from the Indian Ocean to the Pacific. "The South China Sea is an important junction for navigation and an important maritime gateway from China's mainland and nearby islands," declared a *Peking Review* article in late 1975. "The South China Sea Islands are very important geographically as a key link on the arc shipping lane between Kwangchow, Hong Kong, Manila and Singapore."[1] The countless lagoons formed by coral reefs throughout these waters offer ideal shelters for submarines, in particular, and Japan had a small submarine base in the Spratlys during World War II. Later, during the Vietnam war, the United States monitored shipping and submarine movements from a corner of the Pratas Reef area not then under regular Chinese surveillance. The sprawling Macclesfield Bank has also served as a midsea anchorage area for the ships of many nations, including the Soviet Union in past years.

In Chinese eyes, the fact that the Soviet navy must pass through this waterway en route to and from its base in Vladivostok makes control of the South China Sea islands peculiarly desirable in military terms. According to Japanese sources, Peking has maintained some sort of naval bases on Woody (Yung-hsing) Island and Lincoln (Tung) Island in the eastern Paracels since at least 1958. Several patrol vessels and supply ships were sighted there in mid-1971, along with shore-based radar installations; and the flotilla that clashed with the Vietnamese in 1974 numbered seven ships, including Komar-class gunboats equipped with Styx missiles. Like the MIG fighters rushed to the scene of battle, some of these Chinese gunboats could have been based on Hainan, but there appears to be little doubt that expanded naval facilities are being developed on Woody Island, the largest of the Paracel group with an area of 1.1 square miles. In November 1975 Chinese ships and planes conducted much larger exercises than ever before in the South China Sea. Soon after, when three foreign ships were wrecked near the Paracels, Chinese authorities radioed other vessels to keep their distance and Chinese vessels rescued the marooned sailors, demonstrating a much greater readiness for naval action in the Paracel area than rival navies operating there had anticipated. Among other indications of expanding port facilities on Woody Island, official Chinese photos show wharves, breakwaters, a meteorological tower, and an oceanographic station, with patrol boats and motorized fishing craft in the background.[2] Peking has also published photos of new buildings there, one described as the administrative headquarters of the Paracels, the Spratlys, and the Macclesfield Bank, and another as the command post of People's Liberation Army units.[3] A small permanent settlement has reportedly been growing up on the island, reinforcing the irregular, seasonal visits of Chinese fishermen from Hainan who have plied the South China Sea for centuries.

In addition to the military motives underlying Chinese interest in the disputed islands, there have been clear indications that China recognizes the resource potential of the South China Sea. A Japanese visitor to China reported in September 1974 that a Canton television film detailing China's claims to the Paracels showed a glimpse of what appeared to be an onshore drilling rig on Woody Island. British and American intelligence sources have also reported occasional evidence of drilling from a concrete causeway close to shore. When the Thieu regime issued a decree in late 1973 incorporating eleven of the Spratlys into Phuoc Tuy Province, the Chinese Foreign Ministry, protesting the Saigon move, not only said that the South China Sea islands had "always been Chinese territory" but pointedly added that "the natural resources in the seas around them also belong to China." [4] Nevertheless, it would be an oversimplification to suggest that a desire for petroleum riches is the governing factor accounting for China's determined posture with respect to territorial disputes in the South China Sea.

For one thing, the water depths involved pose technological problems that would make unilateral Chinese oil and gas development there on a significant scale much more difficult than in the comparatively shallow waters of the East China Sea, where the water depth averages 200 feet and rarely exceeds 600 feet. Much of the South China Sea consists of steep subbasins that were barely within the reach of Western exploratory drilling technology as of 1977 and appeared likely to remain beyond Western production know-how for at least another decade. In order to explore these deeper waters, let alone produce oil there, China would be forced to compromise its "self-reliance" policy to a degree that is difficult to envisage. Figure 9 shows the limited breadth of the continental shelf running along China's coast from Taiwan to Hainan and thence along the Vietnamese coast, paralleled on the eastern side of the South China Sea by the still more slender strips of shelf adjacent to the Philippine islands of Palawan and Luzon. These narrow shelves tend to drop off abruptly into the abyssal depths below. As for the islands themselves, most of them, situated in midocean, have only small shelves of their own. In the case of the Paracels, the water depth descends suddenly to some 3,000 feet, apparently within fifty miles of some of the islands. Northeast of the Spratlys the depth is three times greater.[5] This geological environment still leaves scope for continuing exploration in the immediate vicinity of the islands in the event that the limited drilling already noted should prove encouraging. For the most part, however, Chinese oil exploration in the South China Sea during the decades immediately ahead is likely to be confined to the continental shelf, especially in the Tonkin Gulf and waters south of Hainan. Chinese access to these areas is not dependent upon control of the South China Sea islands, which prompts us to examine closely and critically the widespread assumption that oil governs Peking's sweeping sea boundary claims.

To put the oil factor into a meaningful perspective, it should be viewed as one element in a more comprehensive Chinese effort to consolidate a position of regional primacy. For example, in making its far-reaching boundary claims, Peking is not necessarily serving notice that it actually intends to undertake oil development within the entire area claimed. Peking may be actuated not only by specific plans or priorities with respect to oil development as such, but also by a desire to corner oil development rights as a bargaining weapon in dealing with littoral states. If Manila or Kuala Lumpur shows a cooperative attitude in matters affecting the

Sino-Soviet rivalry, one of the rewards could well be corresponding Chinese cooperation with respect to the boundaries of offshore oil concessions or the terms of crude imports. At the same time, by adhering to its maximum boundary claims, China would be in a position to choose the best areas for its own deep-water exploration and production should it choose to embark on such a program at some future date. In this connection, it should be emphasized that China has kept its options open, carefully avoiding an explicit definition of what it means by its boundary claims pending the outcome of United Nations deliberations on a Law of the Sea treaty. As of early 1976, Peking had not precluded a "soft" line under which it would treat the waters between the islands it claims as the high seas, and the boundary enclosing them on Chinese maps merely as indications of its claim to the islands themselves. Such a position could be qualified, however, by a Chinese effort to apply the principle of a 200-mile economic zone to inhabited islands. There are endless permutations and adaptations of Law of the Sea concepts that China could invoke to suit shifting political priorities in its relations with the littoral states. Once having extended its control over the islands still in dispute, China would be admirably placed to call the tune not only in sea boundary discussions but also in many issues bearing upon the Sino-Soviet rivalry.

An examination of the offshore concession zones demarcated by Malaysia and the Philippines in relation to Chinese sea boundary claims (Figure 9) suggests that Peking has fixed its claim line with the oil factor clearly in mind. Malaysia, in particular, would have reason for alarm should China continue to insist on its southernmost claim, which cuts a wide swath across the lucrative offshore Shell and Exxon concessions along the northern coast of Sarawak and Sabah. Prior to 1976, all of Malaysia's domestic crude oil needs were covered by Shell's four established offshore fields in Sarawak (Bakan, Baram, West Lutongi, and Baronia) with a production of more than 93,000 barrels a day. Offshore oil production in Sarawak and Sabah could easily reach 500,000 barrels a day following the development of full-scale production at Shell's South Furious and West Erb fields and at Exxon's Semarang and Tembungo fields. Boundary disputes with Peking could conceivably arise with respect to most of these fields, especially Semarang and Tembungo, which are located eighty-five and sixty miles offshore, respectively. Moreover, Sarawak is steadily growing in economic importance for Kuala Lumpur following Shell's offshore gas discovery one hundred miles northwest of Bintulu in waters well within the Chinese claim line. The Bintulu reserves are conservatively estimated at 6 trillion cubic feet, and an $830-million liquefaction plant capable of producing 5 million tons of liquefied gas per year for Japanese consumption is expected to go into production in 1979. Mounting interest in the western part of the South China Sea since this discovery could well lead to the involvement of more foreign companies in potentially disputed areas, even farther offshore, that fall within the Sarawak concession zones demarcated by Kuala Lumpur. The greatest possibility of future disputes between Kuala Lumpur and Peking appears to lie in areas north of Sarawak, where Malaysian concession boundaries extend farther beyond the Chinese claim line than in the case of Sabah and the neighboring British protectorate of Brunei.

In the case of the Philippines, there is less clearly established evidence of offshore petroleum than in Malaysia and Brunei, but several discoveries in 1976 have spotlighted potentially significant boundary conflicts with Peking in disputed areas west of Palawan and Luzon. Since the concession zones allotted by Manila reach much farther into areas claimed by Peking than the offshore areas defined by Kuala Lumpur and Brunei (Figure 9), the long-term danger of serious trouble in these areas is no less than in Malaysia. The most conspicuous of these is the so-called Seafront concession in the Reed Bank area of the Spratlys, where a consortium of Amoco and Swedish interests made a strike in August 1976, prompting a Chinese warning that such "encroachments on Chinese territorial integrity and sovereignty are impermissible." Amoco, which holds a 38.5 percent share in the concession and conducted drilling operations for the consortium utilizing a U.S.-registered drillship, proceeded in direct disregard of State Department advice. A company spokesman said that Chinese claims to the area are "incongruous," and drilling was to be resumed in 1977. Two other potentially explosive concessions in the Spratlys have also been allotted to Philippine companies in the London Reefs and Tizard Bank areas. Philippine President Ferdinand Marcos has pledged to defend oil concessions located up to 200 miles offshore, which covers those so far granted, and is building a new naval base at Ulugan on the west coast of Palawan to be manned with four frigates obtained under the U.S. military aid program. At the same time, Marcos has been seeking improved relations with Peking, including continuing access to crude oil imports from China. Diplomatic exchanges over the oil dispute were under way between the two countries in late 1976, with Manila suggesting a division of the South China Sea under which the Philippines would give up its claims to islands in the Paracel group in return for the Reed Bank and other unspecified portions of the Spratlys. Marcos has also been attempting to enlist fellow members of ASEAN (Association of Southeast Asian Nations) in a joint effort to restrain Chinese ambitions in the South China Sea.

Many geologists regard the area west of Palawan as an extension of offshore areas near Brunei and Sabah, where significant discoveries have already been made, and nine of the eleven offshore service contracts awarded in 1972 were relatively close to the western shores of Palawan. Chevron-Texaco, Phillips, Champlin, and Cities Service are among the foreign companies with a stake in offshore areas close to Palawan that overlap the Chinese claim line. Cities Service, in company with Husky of Canada and Filipino interests, made a potentially significant discovery thirty miles northwest of Palawan in March 1976 just short of the Chinese line. With an outlay of $682 million in foreign exchange for crude oil imports recorded in 1974 and its energy demand growing, Manila is pushing these companies to fulfill their drilling obligations despite the political uncertainties involved.

It should be reiterated at this juncture that the strategic considerations discussed earlier would in themselves give China an overriding motive for its South China Sea claims, even in the absence of the oil factor. It appears to be only a matter of time before Peking seeks to dislodge the military garrisons now maintained by Taiwan and the Philippines in the Spratlys, by either political or military pressures or both. In 1974, by all indications, only the lack of a long enough logistical reach prevented China from moving militarily against South Vietnam in the Spratlys as well as the Paracels. Saigon gave Peking a pretext for action in the Spratlys when it issued its decree annexing eleven islets there and then added insult to injury by occupying six of these immediately following the clashes in the Paracels. Peking apparently decided to limit its action to the Paracels because operations in the Spratlys would have required deployment of its missile-carrying patrol boats more than 750 miles from their home base. The Spratlys lie roughly 400 miles east of Saigon and within easy fighter-plane range of the Philippines, but Peking's nearest airstrips in 1974 were on Hainan. While its TU-16 or IL-28 bombers could have reached the Spratlys, Chinese MIG fighters would not have been able to do so. Once Peking is able to provide fighter cover for operations in the Spratlys, Manila may begin to feel pressures for removal of the two Philippine Marine companies that have manned an airstrip in the eastern Spratlys since late 1975. Similarly, Peking might then seek to displace Taiwan's garrison on Itu Aba Island as well as the Vietnamese forces that have reportedly taken over the Songutay Atoll, previously occupied by the Thieu regime. In these last two cases there are obvious political factors that could moderate the Chinese approach. Peking would play directly into Soviet hands by unnecessarily aggravating tensions with its Communist comrades in Hanoi. While registering formal protests against Taiwan's presence on Itu Aba, Peking has implicitly treated the presence of Nationalist

Chinese as better than no Chinese at all and as preferable, in any case, to Vietnamese or Philippine occupation.

The Historical Argument

China's determination to control the South China Sea is buttressed by strong legal and historical arguments of equal or greater force than those of the other states involved. In certain cases, as previously suggested, it is unclear whether Peking will actually seek to enforce its claims to islands located much closer to other littoral states than to its own territory. This applies particularly to the Tseng-mu Reef, twenty miles off the coast of Sarawak, where China may have maximized its boundary demands mainly to reinforce its case for the Paracels and Spratlys. With respect to most of the South China Sea, however, Peking may well mean what it says. Extreme as its demands may seem, they appear to be justified, in Chinese eyes, by the historical record.

The most widely accepted historical proof that China discovered the Paracels has been found in the thirteenth-century *Chronicles of the Sung Dynasty*, which recounts how the last Sung emperor fled to the islands to escape from a Mongol general. Even more substantial evidence supporting the Chinese case lies in the records of the Ming dynasty navigator Cheng Ho, who sailed regularly past both the Paracels and the Spratlys en route to the Indian Ocean during seven expeditions between 1403 and 1433. It is uncertain whether his forces occupied and surveyed all of the South China Sea islands, as modern Chinese historians have claimed, but he did leave behind a map, drawn in 1430 and still extant, in which the relative positions of the Paracels and Spratlys are clearly indicated. The Chinese names for the two subgroups of the Paracels date back to this period, and a major reef in the Spratly group was named the Cheng Ho Ch'ün Chiao. Subsequently, a Ch'ing dynasty scholar also charted the two groups of islands in another map still extant. It is not clear whether year-round Chinese settlements have ever existed on any of the South China Sea islands. But there is ample evidence of semipermanent seasonal settlements established by Chinese fishermen from Hainan and Kwantung Province at least as long ago as the late fifteenth century.[6]

China has attempted to trace its title to the Paracels as far back as the seventh-century Tang dynasty. In support of this position, Chinese archaeologists claim to have unearthed a ruin on Robert Island (Kan-ch'üan) containing celadon pottery jars characteristic of the Tang period. The timing of the Robert Island find has made it somewhat suspect as propaganda, since it was unveiled in late 1974 soon after the Chinese defeat of South Vietnamese forces in the Paracels.[7] Still, even if it is not authentic, China does have solid evidence of its presence there dating back to the thirteenth century.

In contrast to the extensive Chinese evidence of ancient links to the South China Sea islands, the earliest specific Vietnamese claim goes back only to 1802, when the emperor Gia Long established a company to monopolize exports of the rich fertilizer deposits found in the Paracels. Seeking to reach further back, the now defunct Saigon regime argued vaguely that Vietnamese fishermen had operated in the Paracels area "from time immemorial." The Vietnamese case has stressed that Gia Long formally annexed the archipelago to Vietnam in an 1816 proclamation and that in 1832 his successor, Minh Mang, built a pagoda and a stone monument on an islet known as Banna Rock. Soon afterward, it is said, the annexation was reflected in the inclusion of the Paracels on an 1835 map.[8] China never acknowledged Vietnamese sovereignty over the islands during the nineteenth century, however, and it was a Chinese, rather than Vietnamese, protest that forced the cancellation of a German attempt to survey the Spratlys in 1883. When Vietnam became a French protectorate, a Sino-Vietnamese boundary convention fixed the line of demarcation between China and Vietnam to the west of both the

Paracels and Spratlys. As China sees it, the 1887 boundary convention provided clear international recognition of the South China Sea islands as Chinese territory. The Vietnamese response to this emphasis on the boundary convention has been to cite an 1895 legal controversy over two shipwrecks near the Paracels in which China disclaimed ownership of the islands and thus responsibility for the contested cargoes that were recovered from the wreckage. Chinese writers counter, in turn, by recalling that China finally did follow through on the boundary convention in 1909 when it dispatched a force of three warships and some 170 men, led by a ranking naval commander, Li Chun, who formally took over the islands, putting up flags and markers in addition to conducting a field survey.

In large measure the Vietnamese case has been predicated on the fact that France annexed the Paracels in 1932 and the Spratlys in 1933. But by that time, China argues, the 1887 convention and the Li Chun expedition had sealed the issue. China was quick to react diplomatically, in any case, when France issued its annexation decrees, although it was too preoccupied with the Japanese invasion to prevent the French from occupying the islands briefly in 1937. Japan took over the islands in 1939 and ruled them until V-J Day, only to make way for the rapid reassertion of Chinese control in 1946. Saigon made much of the fact that its representatives at the San Francisco peace conference in 1951 were not challenged when they claimed both the Paracels and Spratlys. But neither Peking nor Taipei was represented at the conference, and Chou En-lai, then foreign minister, made a categorical statement on the eve of the conference reaffirming China's "inviolable sovereignty" over the islands. This declaration was directed not only at Vietnam but also at the vanquished Japanese aggressors who had dared to despoil the islands and who were finally persuaded at San Francisco to renounce all rights in the South China Sea.

According to the canons of international law, China had no more than an "inchoate title" to the islands when

Vietnam came on the scene in 1816 with Gia Long's annexation edict, and it would still have been theoretically possible for Vietnam to have canceled out the Chinese claim with a "final and decisive" assertion of sovereignty. As we have seen, however, there is no evidence that either the Vietnamese emperors or the subsequent French colonial regime ever sent occupation forces to the islands until 1932. On the contrary, France acknowledged the Chinese claim to the Paracels during the period between 1816 and 1932. Vietnamese claims rested on the legal principle of *prescriptive acquisition* as against the principle of *discovery-occupation* asserted by China, but these claims were never backed up by the "continued and undisturbed exercise of sovereignty" associated with the acquisition principle.[9]

As historically rooted nationalist symbols, the South China Sea islands are of concern primarily to the two Chinese regimes and Vietnam. Significantly, Peking and Taipei have both concentrated since 1951 on combating Vietnamese and, to a lesser extent, Philippine moves in the South China Sea rather than on their own mutually conflicting territorial claims. For Peking and Taipei alike, it is a psychological imperative to insist on what is seen as China's rightful position of maritime primacy in the South China Sea, especially against the background of wartime Japanese incursions there; and for Hanoi, there is a comparable compulsion to remind Peking that the days of Vietnamese tributary status are past.

For Hanoi, the urge to escape from dependence on Peking for oil makes the controversy over the islands peculiarly important as the key factor determining how much access it will have to oil-rich offshore areas in pursuing its own petroleum development program. Nevertheless, given the Sino-Soviet rivalry, there would appear to be some possibility of a Peking-Hanoi compromise regarding the islands and the consequent division of contested offshore areas. Peking may be reluctant to let the islands become an entering wedge for expanded Soviet influence. Similarly, there is likely to be considerable scope for compromise sea boundary settlements between Peking and the other littoral states concerned in the context of the Sino-Soviet competition. For Peking, the most important thing is to minimize Soviet military access to the area. For Manila, Kuala Lumpur, Djakarta, and Brunei, Chinese title to any specific island is likely to be much less significant in historical or psychological terms than it would be for Vietnam, and what will count is simply the extent to which Chinese claims impinge on their offshore oil ambitions.

It is relevant to underline here that Peking has avoided direct attacks on Hanoi regarding disputed islands since the defeat of the Thieu regime and has given emphasis to its own claims throughout the South China Sea largely in response to repeated Soviet provocations. When Moscow attacked Peking for its early 1974 action against Thieu's forces in the Paracels, Peking, sensitive to Hanoi's awkward position vis-à-vis Saigon, remained conspicuously silent. When *Red Star* championed Malaysian claims to the Tseng-mu Reef,[10] Peking refrained from a direct reply. When Hanoi later announced the capture of "beloved islands" in the Spratlys following the collapse of the Saigon government,[11] Peking continued to maintain an above-the-battle posture. By November 1975, however, when Moscow stepped up its attacks against Chinese "hegemonism" in the Paracels, Peking had become increasingly fearful of a Vietnamese tilt toward Moscow. It was at this point that the Chinese press presented a detailed reaffirmation of Peking's claims in the South China Sea, complete with the archaeological reports, cited earlier, purporting to verify a Chinese presence in the Paracels as early as the seventh century. Hanoi, while reaffirming its own claims in its maps and official statements, has avoided direct attacks on Peking. Several 1976 attacks referred only to the Spratlys, but it was not clear whether this omission foreshadowed a de facto relinquishment of the Paracels claim.

The Hanoi–Peking
Conflict

Faced with the staggering task of rebuilding a war-ravaged economy, Vietnam has pinned high hopes on the discovery of offshore oil as the key to minimizing its dependence on Soviet and Chinese petroleum imports. Past efforts to find onshore oil with Soviet help have proved only marginally productive, and Hanoi is cautiously searching for politically tolerable ways to enlist Western assistance in its offshore exploration program. As in the case of China, however, Vietnam's "self-reliance" policy has made this process an extremely slow one, marked by painstaking efforts to minimize foreign entanglements and, above all, to avoid any public acknowledgment of foreign help that might tarnish Hanoi's nationalist image.

Even before the end of the Vietnam conflict, Hanoi had taken the first steps toward exploring the Tonkin Gulf and other parts of the continental shelf by entering into technical discussions with Agenzia Generale Italiana Petroli (AGIP), the exploration affiliate of Italy's state-owned Ente Nazionale Idrocarburi (ENI). According to an explicit Agence France-Presse report, later denied, attributed to Egidio Egidi, executive officer of AGIP-Mineraria, Hanoi signed a $10-million agreement with ENI on 18 April 1973 providing for the training of Vietnamese technicians in Milan and for subsequent joint prospecting to be conducted "on an equal footing" over a period of four to five years. The first group of Vietnamese technicians reportedly left for Italy in late 1973.[12] Reading between the lines, it appeared that Hanoi had restricted AGIP to a limited technical assistance program but had left the door open for direct cooperation in exploration and production once Vietnam began to acquire its own corps of technicians. More important, as Egidi saw it, the agreement gave other ENI affiliates an inside track in competing for future sales of petrochemical and refinery plants. That some form of technical assistance agreement had been concluded seemed to be confirmed by extensive Tokyo reports in early 1974 indicating that a Japanese consortium led by the Nissho-Iwai trading combine had talked with both AGIP and North Vietnam concerning a partnership role in any actual oil production. Hanoi itself has never directly acknowledged the existence of any 1973 agreement, and AGIP stated in early 1975 that "we have no arrangements with the government of North Vietnam," [13] suggesting that the plan discussed in 1973 might have been canceled or only partially carried out. In any case, Hanoi has since voiced increasing interest in offshore exploration and on 24 August 1975 announced the creation of a cabinet-level national oil and gas agency. The same French geophysical company employed by China, Compagnie Générale de Géophysique (CGG), was hired in 1976 to conduct seismic surveys in the offshore areas near South Vietnam previously explored under the auspices of the Thieu regime, and negotiations were under way with Norwegian, Japanese, American, and other foreign companies for other forms of help in exploration on politically acceptable terms.

Exploration of the Tonkin Gulf and immediately adjacent areas of the continental shelf cannot be pursued very far in the absence of a sea boundary agreement with Peking. So far as is known, no such agreement has ever been concluded, which may help to explain the apparently slow pace of Vietnamese offshore activity. Similarly, Hanoi would face possible conflicts with Peking when and if it should begin to explore for offshore oil along the southern and eastern coasts of Vietnam, depending on how closely it adheres to the boundary claims made by the Thieu regime. The southeast corner of the large offshore area south of the Mekong Delta that was parceled out to Western and Japanese oil companies by the Thieu regime overlapped with the Chinese claim line west of the Spratlys (Figure 9). The overlap was relatively small in relation to the totality of the concessions but happened to be directly adjacent to the spot where Shell discovered oil in early 1975. With respect to the eastern coast, the

Chinese claim line west of the Spratlys would prevent Vietnam from going much more than one hundred miles beyond the continental shelf in its exploration efforts. This is why the Thieu regime pressed its claims to the Spratlys and Paracels in late 1973 and early 1974.[14] It was after formally proclaiming its annexation of the Spratlys that Saigon authorized Geophysical Services International (GSI) of Houston to begin a seismic survey along the eastern coast with an eye to the allocation of new concessions there. As defined in an official Saigon account of oil activities issued in February 1975 the GSI survey embraced "the northern and eastern continental shelves, including the deep sea area" and was to have set the stage for bidding on new concessions "in the second half of 1975." [15] A well-informed United Nations source later confirmed that the survey had been completed just before the fall of the Thieu government and had included areas beyond the continental shelf. However, it is not clear whether the areas surveyed actually overlapped with the Chinese claim line. Interviewed in November 1974, Trade and Industry Minister Nguyen Duc Cuong, one of Saigon's key oil planners, told me that the survey would only cover waters within twenty miles of the coast or where the water depth did not exceed 600 feet and that concessions would also be limited to close-in coastal areas.

Despite indications of enthusiastic Vietnamese interest in offshore exploration, it appears most unlikely that Hanoi will embark on a collision course with Peking comparable to the course pursued by the Thieu regime. Peking remained firm in its stance toward the disputed islands in late 1975,[16] and Hanoi carefully confined itself to pro forma reaffirmation of its own claims during the first year following its defeat of the Thieu regime.[17] It should be remembered that Thieu was spoiling for a conflict with China over the Paracels and Spratlys in early 1974 as a means of rallying nationalist sentiment and thus bolstering his faltering political position. In its own version of the Paracels clash, Saigon acknowledged that it had been the first to take military action, which it justified by citing Chinese incursions in a Vietnamese-claimed island. Following the Paracels defeat, the government not only sought to mobilize anti-Chinese feeling as such by organizing a hero's welcome for the forty-nine survivors of the clash and by assuming an openly hostile posture toward the Chinese minority in Vietnam as part of an anti-Peking propaganda offensive that persisted for months after the January encounter with Chinese forces. In addition, Saigon attempted to convert anti-Peking feeling into anti-Hanoi feeling by belaboring the Provisional Revolutionary Government in the South (Vietcong) for their failure to join as Vietnamese nationalists in condemning Chinese claims in the Paracels. Even in the face of Saigon's ridicule, the Provisional Revolutionary Government took pains to avoid a direct clash with Peking in its initial response to the Paracels incident, declaring only:

The issues of territorial sovereignty are sacred for each nation concerned. With respect to the territorial boundaries between neighboring countries, there are often disputes left by history which may be very complicated and need careful research and consideration. It is necessary for the concerned countries to consider the question in a spirit of equality, mutual respect, friendship and neighborliness, settling the issue through negotiations. [18]

Nine months after the incident, a visiting Thai journalist quoted the editor of the North Vietnamese daily *Nhan Dan* as saying that "Southeast Asia belongs to the Southeast Asian people. . . . China is not a Southeast Asian country, so China should not have such big territorial waters as it claims." [19] But this was promptly softened by a Hanoi disclaimer that stated:

When the said Thai journalist raised the question of territorial waters of the countries in this region, the Editor in Chief of *Nhan Dan* had this to say: As we are fighting the imperialists, the time is not yet ripe for their concern to that question. Once the imperialists are chased away, negotiations may be held among the countries concerned to settle the said question in the spirit of equality and friendship.[20]

In November 1974 Saigon sought to keep anti-Chinese feeling alive by charging that three Chinese trawlers had been spotted circling the *Ocean Prospector*, the Shell rig then drilling in the South China Sea 195 miles southeast of Vung Tau. A dramatic announcement declared that a naval task force of six ships led by a destroyer escort had rushed to the scene and that a squadron of F-5E jet fighters had been placed on alert.[21] Soon thereafter, my inquiries in Saigon indicated that the number of ships had been exaggerated for political purposes and that the largest vessel involved was a U.S.-supplied Patrol Craft Escort (PCE) vessel, smaller than either a destroyer escort or a Corvette and armed only with forty-millimeter guns and anti-aircraft weapons. Shell's drilling operations manager, Jerome Pfenning, told me that foreign oil companies were provided with PCE protection for their rigs as standard procedure under their agreements with South Vietnam. Moreover, informed sources said that the Communist vessels sighted were not Chinese but North Vietnamese and, in any event, had kept a distance of three miles from the rig. Significantly, the *Ocean Prospector* had just made its first major offshore discovery nearby on 12 October 1974, and Hanoi had lost no time in registering its anxiety, first with a flurry of angry broadcasts and then with a show of force.

Oil and the Fall of Saigon

In evaluating the prospects for Sino-Vietnamese conflict over offshore boundaries, one should distinguish clearly between the advantages that a unified Vietnam can derive from a *modus vivendi* with its giant neighbor and the degree to which the Thieu regime artificially inflated tensions with Peking as part of its frenetic search for ways to survive. At the same time that Thieu was manipulating tensions with China, he also made a conscious attempt to strengthen support for his regime at home and abroad by playing up oil prospects. This may have backfired, however, for the discoveries by Shell and later Mobil during the fall of 1974 and early 1975 could well have helped to trigger the final North Vietnamese offensive against Saigon. Reviewing Hanoi's reaction to Saigon's evolving oil program, one is struck by a steadily growing note of impatience and an undercurrent of anxiety that oil discoveries might become a critical factor in determining the course of the Vietnam struggle. Having found little oil in the North up to that point, Hanoi was acutely sensitive to the accumulating evidence that Vietnam's most promising oil reserves were located in the South.

At first, Hanoi showed a measure of restraint when Saigon opened its first bids for offshore concessions in July 1973, just five months after the conclusion of the Paris peace accords, warning mildly that "only the South Vietnamese people and their genuine and legitimate legal representatives are competent to decide on the utilization of these resources." [22] In February 1974, when plans for a second round of bids on new concessions were announced, a Liberation Radio broadcast to South Vietnam charged that Thieu was "frenziedly sabotaging the Paris agreement." [23] In June 1974, after the new concessions had been formally granted, Hanoi issued an exhaustive policy statement on natural resources, angrily recapitulating all past policy declarations concerning Saigon's oil development and treating the offshore program as decisive evidence that the United States had never intended to live up to the Paris accords:

Although heavily defeated and forced to sign the Paris agreement . . . the Nixon Administration is stubbornly continuing to maintain the warlike Nguyen Van Thieu clique as a tool for realizing its scheme of transforming South Vietnam into a new U.S. colony. . . .

The organ of authority to be elected on the basis of the Paris agreement is the only one qualified and competent to manage and utilize the natural resources in South Vietnam. . . . All acts of taking advantage of the Thieu administration to exploit the natural resources in South Vietnam . . . will be firmly opposed by all the South Vietnamese people.[24]

Once drilling actually started, Hanoi began to point uneasily to the impact that oil discoveries could have as a fresh source of economic support for the Thieu regime. On the one hand, Liberation Radio sought to belittle "Thieu's ballyhoo about finding oil" following premature reports of successful drilling by Shell off Con Dao Island in August. At the same time, the broadcast pointed anxiously to the Thieu regime's hopes that U.S. and other foreign capitalists "will jump in to exploit oil more rapidly and make more investments, so that it can overcome its present economic difficulties. In this respect it has obtained initial precarious results." [25] As drilling proceeded, Communist propagandists intensified their verbal attacks against the "invalid and traitorous" Saigon concessions, while Vietcong sappers blew up oil storage depots operated by Shell and Caltex, nominally in response to supply-depot raids by South Vietnamese forces. Then came the appearance of the three trawlers near the *Ocean Prospector.*

For Saigon, Washington, and Hanoi alike, it was an unmistakably momentous development when on 12 October 1974 Shell made a commercially exploitable strike 214 miles southeast of Vung Tau. Thieu used oil from Shell's "Coconut I" well to light a flame to Vietnamese war dead at the National Day ceremony on 1 November and told the assembled war heroes that the war had been fought "to keep our oil out of Communist hands." "Sadness suddenly yields to pride," reported the official South Vietnam press agency, "in being on the threshold of national prosperity." [26] In a Saigon interview soon after the Shell discovery, U.S. Ambassador Graham Martin was reluctant to concede the importance of oil prospects, insisting that oil riches would merely be "the frosting on the cake" and that Saigon was on the way to prosperity in any case, whether or not oil was found. An American embassy economist was more candid, however, pointing to the critical short-term economic relief that oil discoveries could provide for Saigon even if full-scale production proved to be some years away. Addressing the Saigon Lions Club on 11 November 1974, Bruce A. Gulliver, staff economist in the defense attaché's office, declared:

Vietnam's future as an oil producer appears assured. . . . If current prices for petroleum products are approximately maintained, the long-term benefits to the economy are clear. However, the receipt of proceeds from full development of the fields are many years away. . . . A more promising short-range possibility is the use of verified petroleum reserves in seeking international assistance. For the moment there are few items of collateral more valuable than petroleum—especially for economies such as Japan that are totally dependent on foreign sources. Even those countries with oil may be willing to lend some of their current proceeds either for the purchase of future supplies or to attempt to maintain the price cohesion of the oil cartel by offering to buy up voting allegiance.[27]

Similar hopes were echoed by Trade and Industry Minister Nguyen Duc Cuong, who said in early December that "in a few more months, our oil prospects will be clear to everyone, and we will have no difficulty in borrowing money." One possibility, he said, was Saudi Arabia, "but many others are also interested."

The only dissenting voices in Saigon amid the general euphoria over oil prospects in late 1974 came from a handful of U.S. specialists on North Vietnam who were bothered by Hanoi's increasingly frantic response to the drilling activities by U.S. companies. This anxiety was noted by a *Washington Post* correspondent, who observed in the course of his account of the Shell discovery that "Hanoi-watchers, who have been saying there will probably be no general Communist offensive for at least a year, wonder if North Vietnam's desire to strike when Thieu is weakest may not trigger an offensive sooner now that the presence of oil seems certain." [28] Heatedly dismissing such speculation, Cuong told me that "if we are poor, it gives them just as much of a pretext to attack us, which they have been doing. They are just as likely to make trouble either way." "Whatever we do, we are bound to be criticized by them," echoed Vuong Van Bac, then South Vietnamese foreign minister. Nevertheless, harking back to early 1975, it is difficult to escape the conclusion that Hanoi must have been growing restive in direct proportion to the growth of confidence in Saigon regarding oil prospects.

It was on 17 February 1975 that Mobil made its "White Tiger I" strike, with a flow rate of 2,400 barrels a day, resolving whatever doubts might have previously existed about the dimensions of Saigon's oil potential. By May, Shell was due to finish its scheduled drilling and move into preparations for production. By the end of the year, Cuong had predicted, ten more wells would be drilled, and "the time is not too far off when we will be able to produce five million barrels of oil a day. Say five to ten years." Assessing future prospects in February 1975, the last official review of oil progress issued by the Saigon regime before its demise declared:

It is not overoptimistic to expect that in the next few months through 1975, commercial fields will be declared and exploitation rights will be granted to the relevant concessionaires. With due allowance given for technical preparations and installations, it is very likely that the first commercialization of oil from the continental shelf will materialize if not early then late in 1977.

By the end of the decade, the review added, "it should be within reason to hope for a few hundred million dollars per year in revenue for the government, approaching a billion dollars a year in the early 1980s." [29] As a British journalist skeptically observed at the time, "You could say that the famous 'light at the end of the tunnel' has now been replaced by a naphtha flare." [30]

CHAPTER NINE

East Asia in the Law of the Sea Debate

In assessing the potential for conflict over offshore resources between China and its neighbors, this analysis has deliberately focused on the economic and political forces at work, giving only subsidiary attention to the significant issues of international law involved and the legal principles invoked by the contending states. To some extent, this is because competent studies by Choon-ho Park and others have already considered the legal dimension of offshore disputes in East Asia; [1] to an even greater extent, it is because many of these disputes do not fall into the conventional categories defined by experience in other parts of the world. Geographic and geological factors have combined to pose distinctively difficult problems that may require extralegal solution as part of larger economic and political trade-offs. Given the peculiar characteristics of the East Asian environment, the projected United Nations Law of the Sea treaty is unlikely to produce clear guidelines for the settlement of some of the major conflicts under discussion in these pages. Nevertheless, the mere existence of a treaty could have a major impact on the tactics employed by the disputants, and I will now relate my analysis to the legal questions highlighted by the United Nations debate.

A Special Case

The first attempt to establish internationally agreed criteria governing the jurisdiction of coastal states over offshore resources was the United Nations continental shelf convention adopted at Geneva in 1958. Under this agreement, coastal states have the exclusive right to exploit seabed resources up to a depth of 200 meters "or beyond that limit, to where the depth of the waters admits of the exploitation of the natural resources of the said areas." Where two states lie on opposite sides of a continental shelf, the Geneva convention states, or where they lie adjacent to each other on the same coast, the shelf boundary is to be determined by mutual agreement. If such agreement cannot be reached, the boundary is to be a median line determined by the same base points used by each state in defining its territorial sea, unless another boundary line is justified by "special circumstances." [2]

    The caveat permitting states to claim "special circumstances" touched off an intermittent legal controversy, still unresolved, over precisely what makes this or that island valid or invalid as a base point. Among the many resulting disputes that arose in the East China, Yellow, and South China seas, the most troublesome has proved to be the case of the Senkaku Islands (Tiao-yü T'ai), which is examined later in this chapter. To becloud matters further, the International Court of Justice, interpreting the 1958 Geneva convention in the *North Sea Cases,* held in 1969 that offshore boundaries should be drawn so as to "leave as much as possible to

each party all those parts of the continental shelf that constitute a natural prolongation of its land territory into and under
the sea, without encroachment on the natural prolongation of
the land territory of the other." [3] By emphasizing the natural-
prolongation principle, the Court left it unclear whether the
median-line approach should be applied at all in cases where
a subsea trough divides what would otherwise be a continuous continental shelf between two states. As it happens, there
is just such a subsea divide in the East China Sea. Known as
the Okinawa Trough (Figure 6), it is located to the west of the
Ryukyu Islands, as indicated in chapter 7, and is both deeper
(7,000 feet at some points) and broader (100 miles in places)
than the Norwegian Trough in the North Sea. In the East
Asian context, therefore, the 1969 ruling had momentous implications, providing China with a legal rationale for seeking
jurisdiction over the continental shelf as far as the Okinawa
Trough.

Ironically, less than a year after the Court handed
down its judgment, Taiwan, South Korea, and Japan allocated
their offshore concessions in many of the very areas where
China could assert its new claims under the natural-prolongation doctrine. Battle lines were clearly drawn at that juncture
for a profound legal conflict between Peking and neighboring
capitals, a conflict that was to be closely linked with the developing United Nations Law of the Sea discussions. For all
of the parties concerned, it became increasingly clear that the
terms of a Law of the Sea accord would largely determine
which legal criteria, if any, could provide the basis for a resolution of East Asian offshore disputes, especially since China
and Japan had not signed the 1958 convention.

Turning first to an examination of China's legal posture, one finds a consistent effort to retain the option of
claiming the entire shelf tempered by a carefully hedged
readiness for compromise. Peking reacted sharply in late
1970, when Taipei and Seoul allocated their concessions in
Chinese-claimed areas and when Japan claimed the Senkakus
(Tiao-yü T'ai) as a basis for maximizing its own claims to the

shelf. In doing so, however, Peking stopped short of advancing a precise definition of its own boundary claims, warning in general terms that the resources of the seabed "around" Taiwan, the Tiao-yü T'ai islands, the South China Sea islands, "and of the shallow seas adjacent to other parts of China belong to China, their owner, and we will never permit others to put their fingers on them." [4] Similarly, while supporting Latin American demands for 200-mile zones of maritime jurisdiction in a 1970 United Nations statement, Peking did not specify whether 200 miles represented the full extent of its own offshore claims. [5]

In its first comprehensive policy statement before the United Nations Seabed Committee on 2 April 1972, Peking echoed its earlier reference to the resources of "the shallow seas," a calculated phrase that could be taken to cover the entire continental shelf, but did not take a stand in the debate over how the extent of the shelf (i.e., the location of the continental margin) is to be determined. The 1972 declaration emphasized the "sovereign right" of all coastal countries to

the disposal of their natural resources in their coastal seas, seabed and subsoil. . . . All coastal countries are entitled to determine reasonably the limits of their territorial seas and national jurisdiction according to their geographical conditions, taking into account the needs of their security and national economic interests. [6]

By early 1973 Peking had adapted its position to the developing Law of the Sea debate, giving its support in a Seabed Committee working paper to the principle of an economic zone limited to 200 miles. But this position was accompanied by an all-important qualification that implicitly invoked the North Sea ruling:

By virtue of the principle that the continental shelf is the natural prolongation of the continental territory, a coastal state may reasonably define, according to its specific geographical conditions, the limits of the continental shelf under its exclusive jurisdiction beyond its territorial sea or economic zone. The maximum limits of such continental shelf may be determined among states through consultations.[7]

In view of the fact that the continental shelf extends well beyond 200 miles in the East China Sea, this qualification consititued a major milestone in the evolution of Chinese Law of the Sea policy. Peking had made clear that it would retain its option of claiming the entire shelf irrespective of the terms of a Law of the Sea treaty. At the same time, its impact was softened by the allusion to "consultations" and a related provision that "states adjacent or opposite to each other, the continental shelves of which connect together, shall jointly determine the delimitation of the limits of jurisdiction of the shelves through consultations on an equal footing."

Peking stressed its readiness for regional consultations in another important policy declaration during this period, occasioned by the Japanese–South Korean agreement on joint oil development in February 1974. As we have observed in chapter 6, this declaration once again stopped short of explicitly claiming the shelf. It reaffirmed the natural-prolongation principle, but only to support the Chinese demand for a voice in any delimitation of the shelf, carefully refraining from the definition of a specific Chinese bargaining stance. Similarly, in a UN statement shortly afterward, Peking reiterated vaguely that "all seabed resources in China's coastal sea areas and those off her islands belong to China."[8] It was not until its 13 June 1977 statement, following Japanese ratification of the Japan–South Korea agreement, that Peking staked its claim to the East China Sea shelf in unambiguous terms, terming it "an integral part of the mainland."[9]

By design, Peking has kept its neighbors guessing as to whether it seriously intends to push the natural-prolongation principle or has been using it for bargaining purposes and is actually prepared to enter into median-line agreements. This vagueness not only has the general advantages that go with preserving tactical flexibility but also enables Peking to take into account the special characteristics of each dispute. For it would be one thing to affirm the natural-prolongation principle in the East China Sea—where China, a coastal state, faces Japan, an island state—and quite another where countries are contiguous on the same land mass—as in the cases of China and Korea or China and Vietnam.

Japan has attempted to push its base points for a median line as far to the west as possible on the shelf by claiming the status of "special circumstances" for the Senkaku Islands (Tiao-yü T'ai), in the southern part of the East China Sea, and for two other uninhabited islets, Danjo Gunto and Tori Shima, in the northern part.[10] Both of these are on the "Chinese" side of the Okinawa Trough, however, and in order to win recognition of these claims, Japan would have to prove that it is entitled to "jump" the trough (Figure 6). The argument advanced by Japanese and foreign oil companies with Japanese concessions in the East China Sea is that the trough is a geomorphic *depression* in the shelf but not a geological *breach*. This view has been spelled out most forcibly by Northcutt Ely, a legal adviser to Gulf, which has a partnership arrangement with Teikoku, described earlier, covering more acreage on the shelf than that held by any other company under any concession agreement in East Asia. If the effect of the trough is "to interrupt the continuity of the *legal* continental shelf," Ely has argued, then

Japan's seabed jurisdiction as generated by the Ryukyus terminates somewhere in the vicinity of the Trough. If, on the other hand, the Trough does not interrupt the prolongation of the continental land mass, the seabed between the Ryukyus and the mainland is the common prolongation of both, and Japan's jurisdiction extends past the Trough to the median line as against China.

The lawyers' primary concern in establishing seabed boundaries, Ely has maintained, should be with the topography of the seabed, not with the underlying geology:

Scientists point out that the seabed between the Ryukyu islands and the Asian mainland, when viewed as a relief feature of the earth's surface, can reasonably be considered a physical prolongation of both the mainland and the Ryukyus. The Okinawa Trough is a depression in this prolongation, but does not terminate it.

On this hypothesis, Ely concluded, the eastern terminus of the shelf is the Ryukyu Trench to the east of the Ryukyu Islands in the Pacific, and Japan is thus entitled to jump the Okinawa Trough, since the trough cannot properly be designated as the continental margin.[11]

To the dismay of Gulf and the Japanese oil companies with concessions on the shelf west of the Ryukyus, the Japanese government greatly weakened its position on the natural-prolongation issue by seeking to use the natural-prolongation argument with respect to its own east-coast offshore areas and, above all, by concluding its joint development agreement with South Korea. Covering parts of the shelf closer to Japan and Korea than the Teikoku-Gulf concession (Figure 6), the Seoul-Tokyo agreement was actively encouraged by Caltex, which had concessions from South Korea that overlapped a Japanese concession (Nippon Oil) in which the company also had a major stake. Other Japanese and American companies with overlapping or conflicting Japanese and South Korean concessions had also pushed the agreement. South Korean claims to the contested areas had been based on the natural-prolongation principle, however, and in agreeing to the 1974 compromise, Japan had given a measure of legitimacy to the South Korean legal argument. Japanese critics of the agreement believe that Tokyo should have insisted on a median line drawn on the basis of Danjo Gunto and Tori Shima as base points. By diluting its base-point claims in negotiations with Seoul, these critics hold, Tokyo has undermined its case for jumping the trough in any future Japanese dealings with Peking regarding the shelf.[12]

Politics in Command

In the case of China and Japan, as argued earlier, political factors are likely to moderate a controversy that would be extremely difficult to resolve within a narrowly legal framework. By contrast, in the case of China and the two Koreas, it is the political environment that complicates what might otherwise be legally soluble boundary disputes. If the Korean peninsula were unified, the offshore disputes between China and Korea, as contiguous countries located on the same continental mass, would exemplify boundary problems that exist in many parts of the world and clearly invite the lateral-line solution envisaged in the 1958 Geneva convention, a variant of the median line. Such an approach would not obviate possible arguments over specific base points. Nor is it inconceivable that China might seek to claim primacy on the Yellow Sea shelf for geological reasons differing from those embodied in the natural-prolongation principle. In view of the long-standing Chinese emphasis on the fact that the sediments on the shelf come from Chinese rivers (see chapter 3), some observers fear that China might claim all sedimentary deposits up to the silt line (Figure 1) as Chinese in origin.[13] Such a claim would not have the support in international law provided by the natural-prolongation principle, however, and a median-line approach by China would be more likely in the Yellow Sea but for the fact that Korea is divided.

Peking recognizes North Korea as the only legitimate Korean regime and is strongly committed to the principle of Korean unity. Moreover, from Pyongyang's point of view, as suggested in chapter 6, it is desirable for Peking to paralyze South Korean offshore development by casting doubt on the validity of Seoul's concession boundaries. A median-line agreement with Seoul would appear possible, if at all, only in the event that Pyongyang should give its blessing to such an arrangement as part of an overall North-South accommodation. At the same time, the continued division of Korea might not rule out a median-line agreement between Peking and Pyongyang; and Pyongyang, for its part, has pointedly endorsed the median-line principle in Law of the Sea discussions.[14] Peking left the door open for an agreement with Pyongyang when it referred to joint delimitation of continental shelves "which connect together" in the 1973 Seabed Committee working paper. By the same token, a median-line settlement appears possible between China and Vietnam with respect to adjacent areas of their common continental shelf in the Tonkin Gulf, as distinct from the knottier problems associated with their territorial disputes over the South China Sea islands.

To an even greater extent than in the case of China and Korea, political, rather than legal, factors are likely to determine the future disposition of the extensive offshore boundary disputes between China and Taiwan. The conflict between Peking and Taipei over offshore resources is merely one facet of their unfinished civil war and thus falls outside the purview of the median-line approaches that could be applied to disputes between China and Korea, as separate countries, when and if the struggle for control of a unified Korea is resolved. Indeed, purporting as it does to be an all-Chinese regime, Taipei has a legal posture almost identical with that of Peking in relation to Japan and others. Taiwan's concession boundaries, extending to the edge of the shelf, are implicitly based on the concept of the natural prolongation of the mainland. Unlike Peking, Taipei ratified the 1958 convention, while carefully reserving the right to invoke the natural-prolongation principle in accordance with the 1969 Court ruling.[15] Taipei has also paralleled Peking in its insistence on Chinese, as opposed to Japanese, title to the Senkaku Islands (Tiao-yü T'ai). This underlines the fact that Taipei's presence on the shelf is temporarily useful to Peking in that it helps to preempt Japanese activity there without the necessity for a direct Peking-Tokyo confrontation. In the long run, however, Peking is not likely to tolerate offshore development by a Taipei regime that continues to claim all-Chinese jurisdiction.

Chapter 5 shows why it is critical to recognize the in-appropriateness of a median-line approach in the case of Tai-pei and Peking. The Taiwan concession boundaries were drawn at a time when the United States still recognized Tai-pei as the sole legitimate Chinese regime, and all of these boundaries, to one degree or another, were based on median-line assumptions. Some of the companies concerned chose their concession sites only after elaborate legal judgments as to how a given location would stand up in any formal or in-formal median-line settlement. They were confident that Tai-pei would declare itself a sovereign republic or would at least retain de facto independence with U.S. support for an indefi-nite period. Even before U.S. China policy changed, how-ever, Peking was implacably opposed to the concept of an in-dependent Taiwan. Now the 1972 shift has made it increasingly improbable that any Chinese regime would enter into median-line agreements with Taipei that would imply formal de jure recognition of Taipei as a separate entity.

The application of Law of the Sea principles to an off-shore boundary settlement in the East China Sea would be more likely to come in the context of a Sino-Japanese divi-sion of the shelf in which Taiwan is implicitly or explicitly reduced to the status of a Chinese province and gradually moves into the economic, political, and defense orbit of the mainland. It will be argued later in this chapter that such a scenario would not necessarily mean an automatic end to Taiwan's present economic system, including limited oil ac-tivities conducted independently from the mainland either with or without the collaboration of foreign companies. Its implications would be far-reaching, though, for Taipei would have to surrender its claims to all-Chinese jurisdiction and thus its claims to many of the offshore areas now embraced in its concession agreements. To the extent that Peking should, in fact, prove ready to tolerate economic autonomy for Tai-wan, it might give its de facto blessing to the boundaries of any concessions retained in the immediate vicinity of the

island. But these would not necessarily be based on the median-line concept and would not be enshrined in international agreements.

To summarize, in the event that the projected Law of
the Sea treaty is adopted and incorporates, as expected, a
provision for 200-mile economic zones, it is likely to have
direct applicability in East Asia only in the Yellow Sea, where
Article 71, of the Single Negotiating Text,[16] would provide a
possible basis for a median-line agreement between China
and North Korea or, in the long run, between China and any
unified Korean regime. Median-line agreements should be
based on "equitable principles," according to Article 71, employing the median line "where appropriate" and "taking
into account all relevant circumstances." Since the Yellow Sea
is not wide enough in most places to accommodate 200-mile
economic zones for both China and Korea, the situation there
clearly invites a median-line approach. With respect to the
East China Sea, by contrast, the treaty is likely to be more
open-ended. Should it so desire, China could contend that
Article 71 was meant to apply to cases in which the natural-
prolongation principle can be advanced by more than one
party, but not to a situation in which only one coastal state is
involved. Citing the natural-prolongation principle, Peking
could point to the Okinawa Trough as a "relevant circum-
stance" rendering the median-line principle inappropriate in
the East China Sea. Japan, for its part, could counter by citing
Article 128,[17] which would give islands 200-mile economic
zones. Tokyo then could seek a median line drawn with the
Ryukyu Islands or the Ryukyu Trench as a base point in the
event that its claim to the Senkakus (Tiao-yü T'ai) is not recognized.

As I indicate later in some detail, the treaty is suf-
ficiently elastic in some respects to permit an accommodation
of the seemingly intractable legal issues dividing China and
Japan, especially if political factors continue to be favorable.
In the case of the South China Sea, however, the specificity of

the treaty could aggravate the serious territorial conflicts outlined in chapter 8 if islands are entitled to have 200-mile economic zones and if China continues to assert dominion over
all of the islands it now claims.

Should the treaty be adopted, a period of some years is
likely to elapse before it is clear whether China and each of its
neighbors will ratify it. Even if the treaty were not to acquire
legal force in East Asia, however, its mere existence could
well serve to restrain overt conflict. In this regard, particular
attention should be given to Articles 62 and 71, which en-
vision "provisional" median-line arrangements in cases
where differences exist over the delimitation of the continental shelf.[18]

Beyond the 200-mile
Economic Zone

One of the most sensitive and least understood aspects of the
Law of the Sea debate is the issue of how rights to subsea
resources should be demarcated in those cases—among them
the East China Sea—where the continental shelf extends
beyond the proposed 200-mile economic zone.

Article 64 of the Single Negotiating Text states that the
continental shelf of a coastal state extends "throughout the
natural prolongation of its land territory to the outer edge of
the continental margin" and treats the 200-mile zone as normally applicable only in cases where the shelf is less than 200
miles wide. Article 70 goes on to spell out revenue-sharing
principles governing resource exploitation in areas between
the 200-mile limit and the margin. At the same time, the
projected treaty fails to define the continental margin. The
importance of the "broad shelf" issue has been heightened
not only by the many geological and geomorphic anomalies
affecting the definition of the margin but also by the growing
belief among leading geologists that the deep-water areas

beyond the shelf may be an even greater repository of oil than the shelf itself.

As briefly noted in chapter 7, the continental shelf becomes the continental slope where it drops to depths greater than 600 feet; the slope, in turn, gives way to the continental rise, which merges with the ocean floor; and the margin is generally understood to lie within the area encompassed by the slope and the rise. Precisely where the margin should be defined in relation to the rise, however, is a subject of continuing debate. In geological terms, it was completely arbitrary to adopt 200 miles as the limit for national economic zones, as Hollis Hedberg of Gulf has observed, since

The world's great thicknesses of sediments with promising petroleum prospects are more closely related to the base of the slope than to some fixed distance from the shoreline. . . . Some of the thickest sedimentary deposits known lie just seaward of the base-of-slope line and constitute the so-called continental rises—huge sediment-filled sumps which, although somewhat questionable as regards adequate reservoirs and traps, must certainly have been good generators of petroleum.[19]

In the case of the East China Sea, the issue is complicated by the controversy over whether the Okinawa Trough defines the continental margin or should be considered a mere depression. Geologically speaking, there is a substantial basis for the argument that the trough does not define the margin and that the "true" slope and rise are located in the Ryukyu Trench to the east of the Ryukyus.[20] Nevertheless, in geomorphic terms, the trough, like the Southern California offshore basins, does have the equivalent of a slope and does flatten out in many places, thus serving as a trap where "huge sediment-filled sumps" are believed to have accumulated just as they would in the case of a "true" geological rise.

The promise of deep-water areas has not been widely recognized because the oil industry has generally played down its interest in the slope and the rise in order to avoid stimulating international interest in establishing control over the margin.[21] Moreover, even in industry circles, the margin was often dismissed along with other deep-water areas until relatively recently as being beyond the reach of drilling technology or as offering insufficient petroleum prospects to justify the costs of developing the requisite technology. Increasingly, however, geologists have begun to contrast the long-term potential of deep-water areas with the measurably short-lived potential of known shallow-water reserves. On the basis of the latest research, J. D. Moody of Mobil told the World Petroleum Congress in 1975:

Our opinion of the potential of the world's deep-ocean areas may well undergo significant re-evaluation. . . . The deeper ocean may constitute an area of potentially large underestimating, and our current estimate of ultimate resources could conceivably double. As additional data have become available from studies of the ocean areas in recent years, and as the scramble for additional resources takes on greater urgency, the importance of the continental margins becomes more apparent.[22]

Even without allowing for Soviet and Chinese offshore reserves, Moody said, some 67 percent of the world's undiscovered oil resources "underlie the oceans and their continental margins." In January 1974, eight of the larger Western oil companies had already leased over 300 million acres in deep water, constituting some 15 percent of their worldwide holdings of nearly 2 billion acres. This commitment was not only a reflection of their growing confidence in the geological potential of deep-water areas; it also testified to the continuing progress being made in deep-water drilling and production technology. In 1976 exploratory drilling had reached a depth of 2,150 feet in water 600 feet deep, and production had been undertaken in water depths of more than 400 feet, albeit at inordinate costs that would be commercially viable only in the event of major discoveries. By 1985, according to most oilmen, exploratory drilling and production will both be feasible in water 2,000 to 3,000 feet deep.

As geological interest in deep-water prospects has grown, American oil companies, eyeing U.S. offshore areas, have argued that the continental margin should be demarcated so as to give coastal states control over most of the slope and the rise. This control had been assured under the 1958 Geneva convention, which provided that coastal states were to have jurisdiction over areas of their shelves beyond 600 feet in depth, extending "to where the depth admits of the exploitation of the natural resources of the said areas." By 1969, however, proposals had begun to surface in the United Nations Seabed Committee for a Law of the Sea treaty that seemed likely to depart radically from this approach. Developing countries were pushing a variety of plans for international development of seabed resources, and most of these plans envisaged some form of international jurisdiction over areas beyond 200 miles or, at the very least, beyond some undefined point within the margin in cases where the shelf extended for more than 200 miles. Northcutt Ely, mentioned earlier in relation to the legal concepts underlying Gulf's concession agreements, reflected the views of U.S. deep-sea mining as well as petroleum interests when he warned that proposals for United Nations authority over areas "beyond the limits of national jurisdiction" actually constituted a threat to coastal-state control over the margin. "While these measures purport to deal with the creation of a regime to govern mineral development on the deep sea floor beyond the limits of jurisdiction of the coastal nations," Ely declared, "it has become clear that the urgent issue is the extent to which this future regime shall acquire jurisdiction over the mineral resources of the continetal margin, and receive royalties therefrom." He contended that oil and other resources of the U.S. continental margin "are part of the American mineral estate, the heritage of all our people, and Congress alone has constitutional power to control their use." [23] A committee of the National Petroleum Council echoed this view in 1969, urging specifically that

Where continental rises are developed adjacent to the continental slope, the sediments of these rises will overlap the lower part of the slope so that the true boundary marking the outer limits of the continental block must be drawn to include not only the slope but also the landward portion of the rise.[24]

Five years later, petroleum-industry spokesmen were still lobbying against "any renunciation of existing rights in the U.S. continental margin." [25]

Like other coastal states with broad shelves, the United States has supported proposals designed to retain control over shelf areas beyond 200 miles in the projected Law of the Sea treaty. By agreeing to revenue-sharing in areas between 200 miles and the margin,[26] the United States and kindred broad-shelf powers have persuaded many other countries to accept coastal-state control over these areas. Even if coastal-state control were established up to the margin, however, it would still be necessary to define exactly where the margin is to be demarcated, that is to say, how much of the rise is to be on the landward side of the margin boundary and thus within the province of the coastal state. Proponents of an international seabed authority have been seeking to maximize the revenues of the new agency by getting as much of the rise as possible on the seaward side of the margin. The major oil companies with U.S. offshore interests are searching for formulas to justify keeping as much as possible on the landward side, which would fall beyond the proposed international revenue-sharing arrangements. Hollis Hedberg has proposed a plan that would enable the United States to fix its own line through the creation of an "international boundary zone," beginning at the base of the slope and extending for an internationally accepted distance of at least fifty miles "within which the precise boundary would be drawn by the coastal state itself." Both the base-of-slope line and the margin boundary would be subject to the approval of an international boundary commission. The base of the slope is "not definable sharply enough to serve as the boundary itself," Hedberg has argued; indeed, it is because there are so many "uncertainties in the precise identification of the base of the slope" that it is necessary to have the proposed international zone.[27] Hedberg would do away with the 200-mile economic zone altogether so that the United States, as a broad-shelf power, could have unambiguous control over all offshore areas up to the margin. While endorsing the boundary-zone proposal in most particulars, Robert D. Hodgson, Geographer of the U.S. State Department, has stipulated that it apply only beyond 200 miles, pointing out that many countries with narrow shelves want 200-mile zones to safeguard fishery as well as petroleum resources.[28]

The debate over the demarcation of the margin illustrates that the same rationale will not necessarily serve the interests of American companies at home and abroad. Thus, in the case of U.S. offshore areas, coastal-state control over the precise location of the margin would enable U.S. companies to keep the richest portions of the rise out of the reach of an international seabed authority. By contrast, in the East China Sea, where a coastal state faces an island state, the issue is not between China and an international authority but between China and Japan. If China should insist on the natural-prolongation principle, would it also insist on controlling the slope of the Okinawa Trough, where Gulf, in partnership with Teikoku Oil, and Shell, in partnership with Nishi Nihon, have extensive offshore concessions?[29] Would Japan adjust to such a Chinese posture, or would it stand by the argument, articulated by Ely, that the Okinawa Trough is a depression and that the Ryukyus have their own natural prolongation necessitating a median-line solution?

In order to justify extending its control to the Okinawa Trough, China would not necessarily have to establish that the trough defines the margin. For example, invoking Article 129 of the Single Negotiating Text, Peking could contend that the East China Sea falls in the category of a "semienclosed sea." Invoking Article 130, which vaguely enjoins the parties concerned in such cases to "cooperate with each other in the exercise of their rights and duties," Peking could argue that the East China Sea is susceptible to treatment as a special case that need not be directly governed by the treaty provisions for median lines and 200-mile zones. Peking could then take the position that the natural-prolongation principle gives it a claim up to the trough while still retaining the option of receding to a median-line compromise if it chose to do so.

When one considers the promising oil potential of the Okinawa Trough, it is clear that Japan has a profound stake in reaching a median-line agreement with China. Moreover, when one examines the dimensions of Gulf's sprawling concession arrangements with Teikoku (Figure 6), embracing the trough, it is equally apparent that Gulf's interests largely coincide with those of Japan. Clearly, Gulf would benefit most from a median-line arrangement with China in which Japan would get full control over the trough. Still, there would be considerable scope for Gulf even in the event of a Sino-Japanese compromise giving Japan de facto operational control over areas beyond 200 miles, including the trough, in return for bilateral revenue-sharing with Peking. As for its Taiwan concession, Gulf is gambling that Taipei would retain economic autonomy even if it were to become a nominal province of Peking and would thus keep control over its oil concessions. Under optimal assumptions, in short, Gulf could conceivably end up with some of the most desirable areas on both sides of a Sino-Japanese median line.

CHAPTER TEN

The United States and Chinese Oil

The ubiquitous role of Gulf, discussed in chapter 9, is only the most striking example of a deep and long-standing American interest in the oil potential of the offshore areas adjacent to China. Prior to 1972 this interest was pursued solely through the continental shelf concessions obtained from Japan, Taiwan, and South Korea. Since the Shanghai communiqué, however, the United States has also looked hopefully to China itself as a possible source of crude oil and as a market for American offshore technology. American policy has shifted, accordingly, from the direct encouragement of U.S. concession links with Tokyo, Taipei, and Seoul to a more flexible approach, cautiously attuned to the new U.S. relationship with Peking. The United States has discouraged drilling in disputed waters, as we have seen, but Washington has also refrained from supporting Peking's boundary claims. Given their multifaceted involvement in East Asia, both in exploration and marketing, American oil companies have carefully attempted to keep their options open. Some companies are probing the possibilities in Peking; some are encouraging Tokyo, Taipei, and Seoul in their offshore ambitions; and some are working both sides of the street, waiting to see how the situation develops. The American role could have a major impact on the future of the continental shelf and thus merits special attention as a critical factor defining the policy alternatives that are discussed in chapter 11.

Even prior to the Communist victory in 1949 American oil companies had begun to realize that China might have a significant oil potential, including possible offshore deposits of great magnitude. Former Treasury Secretary John B. Connally, who has taken a keen interest in international oil issues during his intermittent assignments in Washington, recalled in an interview that American oil company geologists "knew there were enormous anomalies in the East China Sea as long as twenty-five years ago, but there was nothing to be done about it once the Communists took over from Chiang." American interest in the East China Sea was not revived until the mid-1960s, Connally said, when worldwide interest in offshore exploration began to intensify. American companies were initially interested in areas near Okinawa, as a territory directly under American control, but their hopes were soon dampened by Japanese demands for the reversion of the Ryukyu Islands. When Connally joined the Nixon cabinet he argued insistently but unsuccessfully that "we should not give Okinawa back for nothing"; he felt that a demand for the retention of offshore rights for American companies "should have been an explicit part of the reversion agreement and would have been a legitimate bargaining posture."

Once Okinawa had been returned to Japan without an "oil clause" in the reversion agreement, American companies became increasingly responsive to overtures from Taiwan, and most of the concessions discussed in chapter 5 had been concluded by mid-1969. At this stage, the U.S. commitment to Taiwan still looked unshakably firm. American officials welcomed U.S. oil concessions along with other investment commitments. To the extent that Washington was mindful of offshore territorial conflicts between Taiwan and Japan, there was a tendency to minimize the potential of these conflicts and to think in terms of joint Tokyo-Taipei development ventures. The possibility of offshore claims by Peking was seriously entertained only by a far-seeing minority of State Department officials. Then came the abrupt Chinese response, recounted in chapter 1, when Gulf deployed its first seismic survey ship near Taiwan. Anxious to avoid giving provocation to Peking during the delicate prelude to Secretary Kissinger's first visit to China in July 1971, the White House announced its policy of neutrality with respect to all offshore territorial disputes in the East China Sea. This policy was consistent with a general posture of neutrality toward offshore disputes in other parts of the world and would have occasioned little surprise had it been restricted to conflicts between Tokyo and Taipei. As I have noted, however, the extension of this policy to the Taipei-Peking boundary conflict was a significant milestone en route to the new China policy signified by the Shanghai communiqué. To the irritation of the oil companies operating in the area, Washington attempted to demonstrate its good faith to Peking in early 1971 by disclaiming responsibility for the protection of any U.S. vessels used in oil exploration and for the fate of American crew members aboard such vessels. This disavowal made it necessary for the companies to employ non-American vessels and crews at a higher cost than would have been involved in operating their own vessels. More important, the U.S. government refused to license the equipment needed to utilize American satellites for navigation purposes, forcing reliance on less accurate shore-based navigation systems.

It is not yet clear whether, or to what extent, American interest in China's oil potential was a significant factor behind the shift in U.S. policy toward Peking in 1971 and 1972. In any event, following Kissinger's mid-1971 preparatory mission to Peking, there is known to have been widespread discussion within the Nixon administration concerning the possibility of Sino-U.S. cooperation in oil development, especially on the continental shelf. State Department associates maintain that Secretary Kissinger did not foresee much scope for such cooperation and was opposed to the inclusion of this issue on the agenda of the Nixon visit. China was still obsessed with fears of foreign designs on its natural resources, Kissinger felt, and overt U.S. interest in oil could jeopardize his efforts to forge an anti-Soviet political alignment. A more widespread view, centered in the Commerce Department, was that the United States should offer to relax some of its trade restrictions on oil-related technology, at the very least, as a means of opening up the Chinese market for U.S. oil equipment and of paving the way for eventual Chinese crude oil exports to the United States. Still another view, identified with Treasury Secretary Connally, was that the United States should try to use its new entrée to Peking to win Chinese tolerance of U.S. oil exploration in concessions leased by Taiwan, South Korea, and Japan.

During a hitherto unreported exchange with Japanese Foreign Minister Takeo Fukuda on 11 November 1971, just three months before Nixon went to China, Connally sought to enlist Japanese cooperation in a joint Washington-Tokyo posture toward Peking with respect to the development of the continental shelf. This was not a White House "feeler" connected with the Nixon visit, Connally said; he was trying the idea out on his personal initiative as a possible element in his plan for a Pacific Basin Economic Community. Oil was to be the centerpiece in a grand political-economic coalition that would balance the power of the European Economic Community. He explained:

We have the offshore technology, and the Japanese have the shipbuilding, that is, the rig-building capacity. I suggested that the rigs could be built in Japan, and that we could work out joint arrangements for exploration on the shelf. I reminded him that Japan would be increasingly vulnerable as a result of its dependence on imported energy. They were in a position to assert a strong title to many of the areas involved. Enough so that someone who could work with the approval of Japan and Taiwan could have gone ahead to develop the oil then. But he was not responsive. Japan had made up its mind by that time not to tangle with the Chinese. They perceived the situation in that regard long before we did. Now Japan is not going to be assertive with respect to the shelf. It's too late to do anything now.

Whether or not Nixon himself raised the issue, there is substantial evidence that White House aides who accompanied the president on his 1972 visit informally signaled U.S. readiness to cooperate with China in both onshore and offshore oil development. One authoritative State Department source declares flatly that the 1972 visit led directly to the sale through Japanese channels of the jack-up mentioned in chapter 4. This rig had sensitive, U.S.-licensed components and could be sold, under the export-control restrictions then in force, only with White House approval. Further limited steps toward relaxing export restrictions were also taken during the eighteen months following the Nixon visit.[1] China was quick to take advantage of the new U.S. policy by placing progressively escalating orders for American oil-field equipment as part of a stepped-up global procurement program. Beginning with orders totaling $10 million in 1973, Chinese purchases of U.S. equipment had reached a total of $110 million by mid-1977. Most of these purchases were to maximize the recovery from existing onshore wells, but a significant portion consisted of components for offshore drilling equipment either to be made in China or to be obtained from nominally non-U.S. sources.

From a Chinese perspective (explained in chapter 4) Peking has already made a major new departure in the implementation of its "self-reliance" policy by systematically acquiring so much American equipment within the short span of five years. From the perspective of Western oilmen, however, Peking has done next to nothing. As the major oil companies see it, the sensible thing would be for the Chinese to turn over their whole exploration and development effort to a consortium of leading foreign companies, who would work on a management-contract basis and be paid in crude. Such a decision would expedite the process by years or even decades, they argue, greatly multiplying the amount of oil available to China both for its political purposes and for financing its industrial imports. At the very least, in this view, Peking should make use of foreign equipment and advisers on a large scale in return for fixed commitments of crude. It would not be enough for Peking to buy U.S. prototypes and make its own equipment, the majors argue; offshore technology, in particular, is too complex to be absorbed within the framework of a rigidly applied "self-reliance" policy.

By playing a role in Chinese oil development and linking this role to payments in crude, the majors would like to obtain at least partial control over the marketing of any Chinese crude exports, thus guarding against the possible emergence of a new power center in the world oil trade beyond the reach of the "Seven Sisters." But Peking would rather make slower progress on its own than become part of a Western-controlled economic nexus.

At intermittent intervals since the Nixon visit, American oilmen and political leaders have gone to China for discussions of possible cooperation that have invariably fallen short of American hopes. In November 1972 the Continental Oil Company (Conoco) agreed to sell offshore seismic survey technology in the hope of getting a foot in the door, but Peking has yet to go beyond this initial arrangement. In May 1973 the Chinese invited Exxon, Gulf, and Mobil officials to tour the Taching fields; two years later, when Exxon's board chairman, J. K. Jamieson, went to Peking on a nominally sightseeing visit as the house guest of U.S. Envoy George Bush, there were still no signs of serious interest on the part of the Chinese in selling crude to American companies. Leading U.S. businessmen who visited China on behalf of the National Council for U.S.-China Trade in November 1973 were politely turned down when they made formal overtures regarding oil, even though they stressed their readiness for arrangements consistent with the "self-reliance" policy. According to the transcript of a joint meeting with Chinese trade officials, a U.S. delegation member, Andrew E. Gibson, president of the Interstate Oil Transport Company and a former assistant secretary of commerce, declared:

Several American firms, some of whom have already made contact with your government, would be interested in knowing what part, if any, they may be able to play in assisting China with onshore and offshore oil and gas exploration, as well as petroleum production and refining facilities, and would welcome the opportunity to enter into discussions towards this end. They fully understand that the Chinese people seek to retain absolute sovereignty and control over their energy resources.[2]

D. C. Burnham, president of Westinghouse Electric, added that oil exports would help to balance China's trade and that "knowledgeable Americans would be glad to help in order to have a source of supply, fully respecting your conditions with respect to exploration and development."[3] But Wang Yao-ting, a top Chinese trade spokesman, was "emphatic," the transcript said, "that China would not explore her natural resources on a joint venture basis with any foreign country. As to advanced equipment and technology, China would likely import it on the basis of necessity."[4] Foreign Trade Minister Li Chiang dismissed the U.S. overtures lightly, stating that China did not yet know the extent of its petroleum reserves and that Chinese technicians "had been directed to drill very deep into the earth in the search for oil, but to refrain from drilling too deep so as not to drill through to the United States."[5] When Commerce Secretary Rogers C. B. Morton visited Peking in June 1975 he reported that he "didn't get anywhere" in his attempts to talk about oil exports to the United States.[6] In a reverse twist, however, Exxon and Gulf have been selling modest quantities of fuel oil to Chinese merchant ships in foreign ports.

To the extent that serious discussions concerning Chinese crude exports have taken place between Peking and

U.S. companies, as in the case of Caltex, they appear to have related not to the American market, but to nearby Asian markets, notably Japan; and as elaborated in chapter 8, Peking may well have had mixed motives in holding such discussions. Peking has been able to use the threat of indirect sales to Japan via American companies as a source of bargaining leverage in its efforts to conclude a long-term oil agreement with Tokyo.[7] Similarly, when Peking held talks with Caltex, Superior, and Socal officials in mid-1977, it was not clear whether this signaled a new readiness for crude sales or a desire to use oil as bait to speed up U.S. action on the Taiwan issue.

The continuing rebuffs to the major oil companies have gradually tempered the enthusiastic U.S. attitude toward Chinese oil development that accompanied the Nixon-Kissinger policy shift of 1972. Initially, in relaxing export restrictions on U.S. equipment, Washington had been confident that significant quantities of Chinese oil would eventually find their way to the United States. By 1977, however, this confidence had begun to wane, and an uneasy policy debate had begun to develop. Would it be in the U.S. interest for China to develop its oil, the skeptics asked, if it did so without large-scale Western cooperation and thus without significant export obligations to the Western majors? Would it be in the U.S. interest if Chinese oil exports went primarily to Japan or if "self-reliance" resulted in little or no exportable surplus at all? Under such circumstances, would it be in the U.S. interest to give renewed encouragement to Tokyo, Taipei, and Seoul in their offshore claims? Or would that jeopardize what were still-fragile U.S. ties with Peking? Torn between conflicting assessments and unsettled by the Chinese succession struggle, Washington temporized; as described by one White House official, the American policy was "not to stand in the way" of Chinese oil development. On the one hand, this ambivalence meant that the United States continued to be more liberal in applying export controls than it had been prior to 1972. In an energy-hungry world, it was argued, there would be some benefit in a successful Chinese oil effort regardless of where the oil was consumed, since more Chinese production would mean less pressure on U.S. sources in the Middle East. On the other hand, as chapter 4 relates, the United States has vacillated as to whether to relax controls on a number of militarily sensitive items needed by the Chinese in their offshore seismic survey program.

To a great extent the debate has been dominated by those in the White House and the State Department who see Sino-U.S. oil issues not in the context of U.S. energy policy but rather as a key element in a foreign policy effort to improve Sino-U.S. relations for anti-Soviet reasons. From this point of view, a liberalized export-control policy has appeared desirable regardless of the Chinese posture toward the Western majors. Similarly, offshore boundary disputes have appeared important not in relation to future oil prospects or in terms of their intrinsic merits but as a potentially disturbing element in Sino-U.S. relations. It would be damaging, in this perspective, if the United States should get caught in the middle of offshore conflicts between Peking and its neighbors, and hence it has been necessary to dissuade U.S. companies from drilling in the more sensitive offshore concessions near Taiwan and South Korea. By the same token, in areas where drilling has not been expected to provoke a Chinese reaction, there is less of a problem and no need for the United States to intervene.

The very nature of the Sino-U.S. impasse over Taiwan has served to deter Washington from developing a clear policy toward the continental shelf as an issue important in its own right. As a logical extension of its effort to maintain simultaneous ties with Peking and Taipei, Washington has deliberately avoided an explicit definition of its position with respect to the validity of Taiwan's oil concessions. The United States has not only adopted a neutral position as between Chinese and Japanese claims to the East China Sea shelf but has also felt constrained from siding with either Peking or Taipei as the legitimate champion of Chinese interests. This

built-in ambiguity in the American legal posture has been accompanied by lingering hopes that the United States may be
able to do business simultaneously with both Peking and
Taipei in the offshore oil realm. Like Connally in 1971, some
oilmen and government officials have continued to argue that
the United States should use its diplomatic links with Peking
to win Chinese approval for U.S. exploration activities in
disputed concession areas leased by Tokyo, Taipei, and
Seoul. In July 1973 George Bush made the rounds of the State
Department and expressed this view repeatedly on behalf of a
group of oil industry friends with interests in the East China
Sea. Then a private citizen between assignments as Republican national chairman and United Nations ambassador, Bush
wanted to know in particular whether it would be safe for
U.S. companies to go ahead with oil exploration in concessions granted by Taiwan. Two years later, a Federal Energy
Agency study suggested that each individual U.S. oil company would have to "choose sides" between Taipei and Peking but took it for granted that U.S. companies, collectively
speaking, would end up helping both Taiwan and the mainland in their offshore development. The U.S. stake in its Taiwan and South Korea concessions, the study observed, is
"substantial." [8] In 1976 presidential aspirant Ronald Reagan
coupled a call for firmness on the Taiwan issue with the confident assertion that the United States could obtain crude oil
from China regardless of its Taiwan stand. [9]

    For most of the U.S. companies with concessions in
areas disputed by China, the refusal of the United States to
back up exploration militarily has made it seem too risky to
proceed with intensive drilling activity in the immediate future. At the same time, the companies concerned are attempting to retain their concessions as long as possible, which has
compelled them to do a minimal level of survey and drilling
work in order to meet their contractual obligations. Their privately expressed hope is that closer Sino-U.S. links might induce Taiwan to seek a more compatible relationship with the
mainland and that such an improved relationship might en-

tail Peking's tolerance of existing U.S. investments and oil-concession agreements with Taipei, possibly in renegotiated form. This hope was reflected in a Ford Foundation report suggesting that even if China "could make good a maximum claim, the only way to gain from it might be to employ the technical skills of the same companies that now hold concessions in the disputed area from Tokyo or Taipei." [10] Significantly, even among these companies, there is a widespread feeling that Peking should be tested from time to time. In the case described in chapter 6, a ranking Gulf official expressed a desire to "force a resolve" between Peking and Seoul by deliberately choosing borderline drilling locations in the Yellow Sea. A similar attitude influenced the Clinton and Superior executives who sought to drill east of Shanghai and the U.S. rig company official who was on the verge of drilling for the Taiwan government oil company near the Senkaku Islands (Tiao-yü-T'ai) in late 1975 until Washington intervened. This feeling reflects, in part, a frankly stated sympathy for Taipei and Seoul as anti-Communist bastions. Essentially, however, it rests on the belief that Peking is dependent on the United States for nuclear protection and might decide to wink at oil exploration activities in disputed areas for the sake of its larger interest in retaining harmonious U.S. ties as a counterweight to Moscow.

It is noteworthy that the United States has attempted to dissuade U.S. companies from drilling only in those areas regarded as most sensitive by Peking and has actually relaxed its pressure on companies seeking to explore in close-in coastal areas near Taiwan and South Korea. As overall relations with Peking have expanded since 1972, Washington has grown progressively less fearful of Chinese military action against U.S. survey vessels and drilling rigs. There have been no fixed guidelines with respect to drilling locations, but U.S. companies have not been discouraged from exploring in the Taiwan Strait and the Yellow Sea so long as their planned drilling sites have been safely beyond what would be likely to constitute a maximum Chinese median-line claim if China should ever agree to the median-line principle. In the case of the Taiwan Strait, as we have seen, Conoco has drilled without U.S. government objection more than 60 miles offshore in areas that could prove to have a closer geological relationship with the mainland than with Taiwan. Gulf also drilled less than 115 miles from Fuchow in mid-1975 with tacit U.S. approval.

After rigid initial opposition to any use of U.S. crewmen, Washington has quietly relaxed its position in the face of persistent company complaints that U.S. specialists are indispensable in many technical jobs. I first learned of this policy change by talking with members of the diving teams and rig crews exploring for U.S. companies in offshore areas near Taiwan in late 1974. Confronted with specific information, the local representatives of the concerned U.S. drilling and diving companies in Taipei and Singapore provided guarded confirmation of the relaxation in U.S. policy, and this, in turn, opened the way for authoritative statements from their head offices. The president of one of the leading drilling companies involved said that most of his rig crews consisted of Americans, but that 40 percent or more were usually non-Americans "so that we can keep operating until replacements come if we ever have to take the Americans off in a hurry." American policy has also been relaxed to permit selective licensing of the equipment needed to use U.S. satellites in navigation. However, the United States has continued to insist on the use of rigs and survey vessels with non-American registry, and Washington intervened decisively in early 1976 to prevent Conoco from using a U.S.-registered drilling rig in the Taiwan Strait.

CHAPTER ELEVEN

Paths to Peace and Development

How does Peking view offshore oil activity by others in areas where it has implicit or explicit claims? What strategy will Peking pursue in asserting its own offshore jurisdiction, and over what time frame? What are possible bases for compromise between Peking and its neighbors that would minimize the danger of regional conflict and permit oil development to go forward?

The View from Peking

The key to answering these questions lies in the assessment made in chapters 1 to 4 that Peking will move systematically offshore but will do so at a pace governed by its ability to absorb foreign technology within the framework of its "self-reliance" policy. In the short run, this assessment suggests, Peking might not make an issue of rival exploration activity in areas where its own ambitions extend; but over time, as its technical capabilities grow, so will its desire to clear the way for its expanding offshore program.

In addition to the economic pressures discussed in chapter 2, there are powerful strategic and political factors that impel China to add an offshore dimension to its oil development. From a military standpoint, there are obvious disadvantages to the excessive concentration of oil facilities in the Dzungarian, Tarim, Tsaidam, and Chiu Chuan basins in the northwest. These basins are not only close to the Soviet border but also consist largely of open desert country that could easily be crossed by armored columns in time of war. The Manchurian fields such as Taching also suffer from the danger of proximity to the Soviet border. To be sure, rigs and production platforms at sea are exposed to naval harassment, but oil facilities on land are more concentrated and thus even more vulnerable, apart from the fact that Soviet naval harassment in international waters would invite the intervention of other powers more directly than the destruction of land-based oil fields. From a political standpoint, offshore activity provides a means of asserting Chinese determination and will in the unresolved controversies over title to the continental shelf, the Senkaku Islands (Tiao-yü T'ai), and the South China Sea islands. These disputes are sensitive focal points of Chinese nationalist sentiment, not only because historical memories are involved, but also because neighboring countries have directly affronted Chinese pride by acting unilaterally in staking out their concessions. In one way or another, China feels compelled to contest this unilateral action; and in Peking's eyes, a singularly appropriate vehicle for such a response is an offshore program of its own that lays down an indirect challenge to rival territorial claims without the disadvantages of a frontal collision.

Military intervention is one way of asserting Chinese claims, as the 1974 takeover of the Paracels from the Thieu regime vividly demonstrated. The continuing danger of military conflict over oil in East Asia has also been underlined by the recurring incidents, cited earlier, of Chinese military surveillance near Taiwan and South Korea and of Japanese naval reconnaissance near the Senkakus (Tiao-yü T'ai). The Paracels case was a distinctive one, however, involving a territorial issue that went beyond oil claims as such and occurring within the context of the Vietnam war. If there is a significant danger that China will resort to military force over oil claims in the future, it would appear to be greatest in the cases of Taiwan and South Korea, since Peking regards both of the regimes concerned, like the Thieu regime, as illegitimate and impermanent. By contrast, where China has been moving toward improved political and economic relations with the country concerned, as in the case of Japan, the use of military force in oil-related disputes would appear likely only as the harbinger of an overall return to a hard line. Diplomatic moves to settle offshore boundary conflicts would be a logical extension of the soft-line policy pursued by China in Asia since the Shanghai communiqué, but Peking appears reluctant to dignify its offshore disputes by agreeing to formal negotiations. What Peking has apparently decided to do, therefore, is to postpone definitive action until its own offshore capabilities are more advanced. As China acquires rigs capable of operating in deeper waters, it could well deploy them selectively in disputed areas, drilling symbolic wildcat wells in preference to sending gunboats. In such a war of nerves, others might well decide to back off in some of the key areas concerned rather than risk a military riposte. Already, in the Superior incident, the United States has shown conspicuous deference to incipient Chinese claims that have not even been formally spelled out yet.

In assessing the future of the contested offshore areas, it is often said that their development will be delayed until sea boundary settlements can be reached, as if the boundary issues themselves were the determining factors involved. This is an excessively legalistic way of looking at problems that are likely to be resolved in a relatively inchoate fashion over a protracted period of time. It would be more meaningful to say that their development will come closer to realization to the extent that China is able to participate in planning and executing the process on a basis of greater technological parity. This view is supported by the element of studied vagueness in the Chinese stand on Law of the Sea issues as they apply to Asia. By avoiding a precise definition of its attitude toward possible boundary settlements, Peking helps to paralyze offshore oil and gas production until it is prepared to play its hand.

Significantly, China makes a careful distinction between exploration and production. Chou En-lai did so on the record in a discussion of Japanese claims to the Senkaku Islands (Tiao-yü T'ai) with a visiting group of overseas Chinese in early 1973. When a questioner pointed to Japan's interest in the possible oil deposits there, Chou replied that "if they explore for oil, we will let them proceed, but exploitation is different and is absolutely forbidden. If they start drilling, we will intervene and stop them." [1] Similarly, in Peking's denunciation of the Japan–South Korea oil agreement in February 1974, the operative passage warned that if the two countries "arbitrarily carry out *development* activities in this area, they must bear full responsibility for all the consequences arising therefrom." (Italics added.) [2] As informally expressed, the Chinese attitude is that Peking has nothing to lose by letting foreign exploration activities proceed up to the very brink of actual development because the findings will be useful for China's own oil program. Once Chinese offshore claims are asserted, it is said, Peking can then seek to obtain the data resulting from these exploration efforts by offering compensation to the foreign interests involved. This attitude could produce a deceptively mild Chinese approach to offshore controversies during the next several years. Given the apparent limitations of the seismic survey program it is developing, Peking is unlikely to complete a preliminary assessment of its vast continental shelf before 1979. By that time, however, Peking will have acquired the first of its imported deep-water rigs, and additional, domestically made rigs may also have been completed. A more assertive Chinese stance would then be a real possibility, especially if there should be dramatic oil discoveries or embarrassing "blowouts" leading to pollution. The offshore issue could readily become entangled with unresolved power struggles in China and interlocking controversies over "self-reliance" and relations with the United States.

The linkage between offshore oil development and the economic strength of a non-Communist Taiwan or South Korea makes these cases inherently more sensitive than the straightforward boundary issues between China and its Communist neighbors, North Korea and North Vietnam, or between China and Japan. The future of Taiwan, in particular, would appear to be closely linked with the offshore issue. Peking would not necessarily oppose offshore development by Taiwan and its foreign collaborators in the context of a gradual Peking–Taipei accommodation. But by the same token, the use of offshore riches to move toward a sovereign Taiwan could trigger a hardening of the Chinese posture.

In Chinese eyes, the oil potential of the Taiwan Basin has increasingly become a focus of national hopes for a rapid breakthrough in industrialization and the early achievement of superpower status. This was strikingly demonstrated in an impassioned defense of Chinese shelf rights by a Hong Kong monthly following the initial Japanese claim to the Senkaku Islands (Tiao-yü T'ai) in early 1971. Complaining that "our shortage of energy resources has been a fatal weakness which makes it impossible for us to compete with the United States and the Soviet Union," *Ming Pao* hailed Western estimates that the Taiwan Basin alone held reserves of some 80 billion barrels (10.7 billion tons). A dramatic chart was presented (Figure 12) to show that

Even without the Yellow Sea, the Taiwan Strait and the Gulf of Tonkin, the oil deposits in the Taiwan Basin enable us to surpass the U.S. and the Soviet Union in oil resources. This demonstrates the importance of the Taiwan problem. We are convinced that the current political situation of China is a passing phenomenon and that China is bound to reunite. When the day of reunification comes, China's industries will be nourished beyond measure by the development of the Taiwan Basin and other offshore oil.[3]

While many Chinese officials are reluctant to discuss the oil aspect of the Taiwan issue publicly, Hsu Ting-mei, a spokesman for the China Resources Company, explained in a Hong Kong interview:

The present situation is temporary, and we approach this matter as part of the overall problem of Taiwan. At present, Taiwan has to import oil from Arabia, but even if they found oil and became rich, it wouldn't make any difference. Rich or poor, Taiwan will eventually come to China. We would not permit this to make any difference.

On the basis of other conversations with Chinese officials in Peking, Tokyo, and New York, it is clear that one of the most significant factors affecting the Chinese posture toward expanded petroleum production by Taiwan would be whether any oil and gas produced in offshore areas is tied directly into the welfare and development of Taiwan or is exported. Since China expects Taiwan to be under mainland control sooner or later, any contribution to its development is regarded, in principle, as desirable. By contrast, export earnings from oil could be viewed as a new source of foreign exchange for armaments and other trappings of state power serving to prolong the life of the Kuomintang regime.

Related to this factor is the issue of whether Taiwan continues to rely on foreign companies or upgrades its own independent capabilities in oil exploration. An analysis of Chinese attacks on the Taiwan oil program shows that Peking does not criticize Taipei solely for its territorial usurpation. An even stronger theme in Chinese statements is the importance of "self-reliance." Reacting to the Conoco gas discovery in August 1974 in a broadcast beamed from Fukien to Tai-wan, Peking charged that "the so-called Chinese Petroleum Corporation of Taiwan has actually sold out the natural resources and rights and interests of the motherland in the name of China." The broadcast added that the petroleum industry on the mainland had developed on the basis of "self-reliance," whereas the "Chiang gang's petroleum plundering plan in Taiwan is comparable to an unfilial son who stole the family fortune by colluding with outsiders." [4] Non-Communist Chinese publications in Hong Kong have also frequently attacked the terms of Taiwan's 1970–71 oil agreements as excessively hospitable to foreign oil companies. "Half of the profits will go to the United States," charged a leading monthly. "Why not wait until you can train your own technical experts and raise the capital to develop the oil yourselves? Why let others exploit you?" [5]

It would be a departure from the main line of this analysis to attempt an explanation here of the deeply rooted "self-reliance" commitment that actuates Peking not only in its oil policy but in all fields of economic development. [6] Suffice it to say that China's underlying nationalist objective since the days of Sun Yat-sen has been to balance its relative power position vis-à-vis that of the West by building an independent economic base. In nationalist terms, the goal is not economic development or economic welfare, as such, but rather the economic independence needed to achieve a more equitable status in global power relationships. Thus, to maximize oil production at the price of becoming dependent on Western companies would not represent economic "growth" at all, in Peking's eyes, and could actually prove to be retrogressive, especially if it entailed more of an export commitment than would be consistent with the overall national drive for "self-reliance" in other economic spheres. For this reason, there would appear to be little basis for the belief expressed by some observers that Peking would sacrifice its "self-reliance" policy as a quid pro quo for an unequivocal U.S. anti-Soviet alignment with Peking and a U.S. rupture in relations with

Taiwan.[7] Normalization of relations with Peking might enable U.S. firms to compete more effectively for oil equipment sales, but the "self-reliance" policy is not likely to be fundamentally altered.

With respect to Taiwan, the "self-reliance" issue has special psychological significance against the background of the Chinese civil war. Peking has long regarded the Kuomintang leaders as collaborationist "puppets" who have leaned on the United States and Japan to compensate for their lack of domestic support. The oil concessions granted by Taipei are viewed as illicit payoffs primarily designed to prolong U.S. backing for the independence of Taiwan at the expense of broader Chinese nationalist objectives. Given this imagery, the future status of these concessions is likely to be directly linked with the course of Taipei-Peking relations. Should Taipei move toward an accommodation, Peking might well temporarily mute its objections to continued exploration, and the issue would not come to a head unless further significant discoveries were made. Conversely, should Taipei declare itself a sovereign republic, Peking might even seek to thwart exploration activities. Should Taipei break loose from its concession agreements and develop its own form of "self-reliance," Taiwan's oil program could conceivably be viewed by Peking as a bona fide "Chinese" effort that could someday become an auxiliary of the mainland's own exploration program. This last eventuality has been privately discussed with the author by certain Taiwan officials. But it would presuppose a political accommodation in which Taipei would acknowledge Peking's suzerainty and accept some form of provincial status in the hope of preserving economic autonomy.

The pros and cons of whether such an accommodation is desirable—or of whether an accommodation could, in reality, result in anything short of total absorption—encompass a variety of considerations beyond the scope of this work. To the extent that a peaceful accommodation short of absorption is possible, however, it would clearly require that Taipei resile from its claims to jurisdiction over the mainland; and

this, in turn, underlines the sensitivity of Taiwan's oil concessions in the East China Sea. Peking's attitude toward future offshore oil activity by Taipei is certain to be influenced
by where drilling occurs and what status for Taipei each location implies: an all-Chinese regime with jurisdiction over the
mainland, a sovereign Taiwan coequal with Peking in Law of
the Sea terms, or provincial status within a Chinese framework. In the Superior incident, drilling was scheduled in an
area that could only be claimed, juridically, by a government
purporting to rule the mainland. The United States reacted
unambiguously in that case but has avoided further definition of the proper limits of Taiwan's offshore jurisdiction.

Taiwan: The Pivotal Issue

As indicated in chapter 9, the application of Law of the Sea
principles to the East China Sea, including the possibility of
a median-line agreement between China and Japan, is
seriously complicated by the continuing impasse between
Peking and Taipei. For Japan, therefore, the linkage between
offshore resource development and the Taiwan issue has a
special meaning. A change in the status of Taiwan could open
the way for a major expansion of Japanese offshore activity
that could prove even more economically advantageous to
Tokyo than the present situation. To be sure, the status quo
is extremely convenient for Japan, since American policy toward Taipei and Peking makes it possible for Tokyo to have
the best of both worlds. Unlike Washington, Tokyo enjoys
the rewards accruing from formal recognition of Peking while
profiting at the same time from the secure trade and investment environment provided by a continuing U.S. diplomatic
and military commitment to Taipei. Looking ahead, however,
many Japanese recognize that the present situation is unlikely
to go on indefinitely and that a new status for Taiwan would
have significant advantages for Japan if it facilitated a Peking-

Tokyo understanding with respect to the delimitation of the East China Sea shelf.

In the scenario most often suggested by Japanese business and government leaders, a U.S. normalization of relations with Peking leads to Taipei's gradual diplomatic isolation and a newly conciliatory posture toward the mainland. This change on the part of Taipei is paralleled by the successful conclusion of a Sino-Japanese friendship treaty and an increasing Japanese "tilt" toward Peking in the Tokyo-Peking-Moscow triangle. Japan and China then reach a median-line agreement in which a greatly weakened Taiwan is tacitly treated as a province of China and is induced (by economic rewards from Washington, Tokyo, and Peking) to acquiesce in the new dispensation by quietly phasing out most of its offshore concessions to the north of the island. Taipei continues to enjoy economic autonomy, for the most part, even retaining many of its foreign trade and investment links, but it must increasingly defer to the Communist regime as the only legitimate government of China and must make adjustments in its foreign economic policies, accordingly, to avoid conflict with those of the mainland. In some spheres, including oil development, Peking demands revenue-sharing arrangements and "interprovincial" trade agreements.

What most Japanese officials appear to have in mind is a median-line agreement that would permit oil development to go forward regardless of how the issue of title to the Senkaku Islands (Tiao-yü T'ai) is resolved. Even if the Japanese claim to the Senkakus (Tiao-yü T'ai) were to be ignored in demarcating a median line, observed Akinobu Tsumuru, research director of the Japan Petroleum Development Corporation, a line drawn on the basis of the Ryukyu Islands as base points would still give Japan substantial scope for oil development in most of the areas covered in its five concessions (Figure 11). The Senkakus (Tiao-yü T'ai) would fall on the Japanese side of the line, but the problem of oil development could be separated from the issue of sovereignty. One school of thought in Japan holds that the islands and a specified area around them should be treated as a disputed zone, with any oil development there regulated by cooperative arrangements. The actual task of development could be entrusted to one of the two countries under a profit-sharing formula, or the two could establish a joint venture. Another proposal would apply the Persian Gulf "enclave" model, generally identified with the Saudi Arabia–Iran agreement governing the status of the Al Farisiyah and Al 'Arabiyah islands.[8] Such an arrangement would give sovereignty over the Senkakus (Tiao-yü T'ai) to one of the countries but would deny their use as base points in fixing a median line. Chinese sovereignty over the islands and a twelve-mile territorial sea around them could be coupled with a median-line solution that would give the oil-rich surrounding area to Japan. Alternatively, Japan could be given sovereignty, but under different median-line assumptions that would give China most of the shelf, or sovereignty for either could be linked with a joint development approach in a specified zone exceeding twelve miles.

"At present, the idea of a cooperative venture in any form does not appear to be acceptable to Peking," Masao Sakisaka, president of the Institute of Energy Economics, stated, "but in five or ten years, who can say? First of all, it would be necessary for the Taiwan issue to be resolved, in any case, before negotiations of any kind with respect to the Senkakus can be imagined."

Although Chinese officials are reticent about discussing their attitude toward the future of the East China Sea, some have informally stressed that a basis for compromise can be found if Japan abandons "legalistic" or "historical" arguments and treats the problem on a more expedient level as one of regional economic cooperation. The clear implication is that Japan must accept Chinese sovereignty over the islands as a prerequisite to any discussion of cooperative oil arrangements. An article by a Western scholar in the Japanese journal *Pacific Community* suggested that this approach would be worth considering, citing the precedent of the 1925 Svalbard agreement by which Norway gained sovereignty over the Spitsbergen archipelago in return for military neutralization and a pledge to share future resource exploitation with the Soviet Union and other signatories.[9] But the idea of acceding to Chinese sovereignty over the islands has not been seriously considered in Japan, and it is widely assumed there that the issue of sovereignty will be deferred indefinitely by mutual agreement, with de facto control hopefully falling to Japan as part of a favorable sea boundary settlement in which the median line is located to the west of the islands. Japanese hopes for a generous Chinese attitude on the location of a median line were enhanced by a 1975 Peking-Tokyo agreement on fishing zones (Figure 11), although the issues involved are different and there is no evidence to support the belief that Peking sees the fishing agreement as the precursor of an agreement governing petroleum development.

It is difficult to foresee how Japan would react if Peking should insist on the natural-prolongation principle and refuse to accept a median line, especially if such an attitude were accompanied by an intransigent posture on the Senkaku (Tiao-yü T'ai) issue as one aspect of a shift to a hard line on Taiwan. The danger of a collision cannot be ruled out, for Peking has consistently based its claim to the islands on the argument that they are an integral part of Taiwan. Former Deputy Premier Teng Hsiao-ping explicitly warned that the struggle for the Senkakus (Tiao-yü T'ai) would be "protracted," explaining:

This question was evaded and put aside by both sides in the establishment of diplomatic relations with Japan. We will never give up this Chinese territory but neither will Japan surrender it. This presents a problem. The priority given to the movement may be high at one time and low at another. For example, it was high when Japan wanted to occupy it in the past, but it is low when this question is not brought up. . . . There will be work to do when the question of whether or not to reinforce the movement for guarding Tiao-yü T'ai is linked up with the question of Taiwan.[10]

The possible ramifications of the Sino-Japanese dispute over the East China Sea shelf have been highlighted by an intermittent Japanese effort to enlist American support for an advantageous median-line settlement in return for the promise of Japanese-American collaboration in offshore development. According to Japanese oilmen and government officials interviewed in the course of my study, this effort has been pursued through both private and diplomatic channels. Minoru Kawamoto, managing director of Kyushu Oil Development, which had a concession in the Senkaku (Tiao-yü T'ai) area until 1975, expressed the hope that Secretary Kissinger would be "very firm and very insistent in pressing China to accept a median line." At the very least, Kawamoto said, the United States should push China to accept a cooperative "unitization" formula in the Senkaku (Tiao-yü T'ai) area similar to that regularly used in the United States when competing companies have concessions impinging on the same geological structure. Masao Araki, executive vice-president of the Toyo Oil Development Corporation, which now holds Japanese concession rights in the same area, argued that a U.S. role would be necessary for the success of any Sino-Japanese negotiations over the Senkakus (Tiao-yü T'ai), if only to help in pacifying Taiwan. Among other things, Araki explained, the United States would have to see that Taiwan continued to get oil on desirable terms from sources other than the Senkakus (Tiao-yü T'ai), since Taipei would probably not be a direct party to any Tokyo-Peking agreement. Yutaka Ikebe, managing director of the semigovernmental Japan Petroleum Exploration Company, emphasized that American firms would be invited by Japan to help develop the East China Sea shelf and that Washington thus has a direct stake in promoting a median-line settlement. Japan needs American technological help, he added, and would, therefore, be prepared to give American firms a "major" partnership role under overall Japanese control.

Conflict or Cooperation?

The search for bases of compromise on offshore boundary issues between Peking and its neighbors is integrally related to the outcome of the controversy in the United States over how and when to proceed with the recognition of Peking as the sole legitimate government of China and the withdrawal of recognition of Taipei. Given the complexity of this controversy and the possibility of considerable delay in the normalization of relations with China by the United States and the other countries that continue to recognize Taiwan, these countries could take a useful interim step to reduce the danger of oil-related conflict by clearly dissociating themselves from offshore claims by Taipei that implicitly represent an assertion of all-Chinese jurisdiction. In so doing, they would also be taking a significant new step toward the de facto recognition of Peking as a prelude to de jure normalization. Such a move will be a necessary accompaniment to normalization, in any case, even if it is not made in the interim.

As a province of China, Taiwan would not be likely to have responsibility for offshore exploration beyond close-in areas immediately adjacent to its shores. Both before and after normalization, therefore, the United States should make clear that it does not recognize Taipei's title to concession zones located more than fifty miles, or some such distance, from the island. To maintain the present deliberately ambiguous policy as to where drilling activities would be legitimate could well be interpreted by Peking—and Taipei—as indicative of a lingering U.S. attachment to a "two China" policy or to the idea of a sovereign Taiwan coequal with the mainland in international law. Such ambiguity adds to the danger that Taiwan might declare itself an independent republic in response to a U.S. normalization move. Conversely, a clearly defined U.S. posture would encourage Taipei to relinquish its offshore claims, promoting a climate of accommodation with the mainland in which the risks of a U.S. military withdrawal from Taiwan would be reduced.

By dissociating itself from Taipei's offshore claims, Washington would in effect be treating Peking as the sole legitimate champion of Chinese claims in the Sino-Japanese shelf controversy. That need not mean taking sides, though, and Washington could best contribute to a peaceful resolution of the dispute by carefully avoiding identification with either Peking or Tokyo. A U.S. posture of noninvolvement would be consistent with the cautious approach marking the U.S. return of the Senkakus (Tiao-yü T'ai) to Japan in 1972 as part of the Ryukyu reversion agreement. Washington specifically stated that the inclusion of the islands was not meant to signify support for Japan in its territorial dispute with China. Having acquired the islands from Japan by conquest, they were being returned to Japanese administration without reference to the preexisting title controversy.[11]

It is necessary to emphasize the continuing importance of a detached U.S. stance in view of the many expedient arguments that are often advanced in favor of U.S. intervention in the shelf dispute. Advocates of a Sino-U.S. alliance against the Soviet Union seek to make a case for supporting Peking's shelf claims as an element of a broader strategic partnership. Those who give priority to the Japan-U.S. relationship point to Japan's importance as an ally in economic as well as military terms, including the heavy financial stake of U.S. and other Western oil companies with shelf interests in a settlement favorable to Japan. Still others see the United States in the role of a peacemaker, capable of orchestrating an equitable solution that would assure a concerted Sino-U.S.-Japanese response to Soviet power. All of these supporters of an interventionist U.S. role underestimate the complexity of the dispute and, above all, its interstitial relationship with the larger fabric of Sino-Japanese relations discussed in chapter 7.

To the extent that compromise is possible, it is most

likely to come as part of a growing network of political and economic trade-offs between Tokyo and Peking reflecting a heightened sense of regional mutuality of interest in relation to other powers. Thus, with respect to the Senkakus (Tiao-yü T'ai), the most promising possibilities for a settlement would require an application of the "enclave" model, but each of the specific proposals suggested earlier would involve real or perceived losses for one side or the other in a calculus restricted by the confines of the shelf dispute as such. These losses would have to be offset by gains in other spheres of their relationship in order for a compromise to emerge. Any attempt to address the dispute in a vacuum, as it were, by means of external support for any one of these proposals on its merits would be likely to appear partisan in the eyes of Tokyo or Peking or possibly both. Such an attempt could easily evoke suspicions of collusion and could complicate what is likely to be a long and tortuous Sino-Japanese bargaining process under the best of circumstances.

In general, there is relatively little scope for constructive external intervention by non-Asian powers or international agencies in the resolution of offshore boundary disputes between China and neighboring countries, especially since Peking has, so far, rejected the dispute-settlement machinery envisaged in the Law of the Sea negotiations. Should global dispute-settlement machinery ever be established, it would clearly be desirable to extend its application to Asia. In the final analysis, however, the outcome of offshore boundary disputes involving China is likely to be governed by political trends in the area; and once this is recognized, it is readily apparent that direct or indirect external involvement in these disputes can have negative as well as positive effects, often by deliberate intent. One conspicuous example of politically motivated intervention by a foreign government has been the encouragement given by the Soviet Union to Vietnamese, Malaysian, and Philippine claims in the South China Sea. Such intervention accentuates the strategic aspect of these disputes and makes it more difficult than it would otherwise be for Peking to scale down its far-reaching claims. As suggested in chapter 8, Peking may well have maximized its claims, in the first place, partly for anti-Soviet bargaining purposes, and the prospects for compromise would patently be enhanced if these disputes could be insulated from the Sino-Soviet rivalry.

So far, at least, other powers have not taken sides directly in China-related boundary controversies, but the role of private Western companies discussed throughout this book often constitutes an indirect form of intervention. The danger of oil-related conflicts would be significantly reduced by a more comprehensive effort on the part of the governments concerned to discourage the involvement of private Western companies in areas where sensitive disputes exist,[12] especially in areas where an unresolved civil conflict is in progress. As the Taiwan case has demonstrated, foreign companies are serving political as well as economic ends when their involvement strengthens one party in a civil conflict vis-à-vis its rival. While that could be said of all foreign investment in divided countries, the fact that offshore oil concessions involve the physical extension of territorial jurisdiction distinguishes these cases from other types of investment.

In the case of Korea, as suggested in chapter 6, the potential for conflict over offshore boundaries arises not only in relation to China, as such, but also in the larger context of the interplay between Peking, Seoul, and Pyongyang. The North challenges the legitimacy of South Korean offshore concessions, prompting Peking, as its ally, to help in paralyzing Seoul's offshore program by asserting Chinese title claims and by refusing to deal with the Park regime as a bona fide Korean government. The United States has warned U.S. companies against drilling in mid-Yellow Sea areas likely to be contested by China in any median-line negotiations. But it has yet to question exploration closer to the South Korean coast, even though Peking is not committed to the median-line principle and Seoul's jurisdiction over its concession areas is challenged by Peking and Pyongyang alike. Japan, too, has disregarded the civil war aspect of the Korean situation by entering into its joint oil development agreement with Seoul, notwithstanding Peking's alignment with Pyongyang and its continuing protests against the accord.

At the very least, it would appear desirable for the governments concerned to make clear to private companies that the risks involved in assisting the South Korean offshore program relate not only to potential conflicts with Peking but also to possible conflicts with North Korea. As shown in chapter 8, the North Vietnamese decision to open a full-scale offensive against Saigon in 1975 might well have been accelerated by oil discoveries that made the Thieu regime more credit-worthy and gave it an economic basis for survival despite its political weaknesses. Similarly, Pyongyang is extremely sensitive to the impact that oil discoveries could have in stabilizing Seoul economically. To put the situation in perspective, a direct repetition of the Vietnamese experience is most unlikely, since Pyongyang is not nearly as strong in relation to Seoul as Hanoi was to Saigon, either militarily, politically, or economically. Nevertheless, the oil issue could become a significant tension point in Korea, underlining the urgency of concerted efforts on the part of Washington, Tokyo, Peking, and Moscow to defuse the North-South rivalry. Precisely because oil is such a strategic factor in the Korean equation, significant oil development efforts would be extremely explosive in areas involving Korea unless and until these efforts can be detached from the North-South struggle through some form of cooperative approach in which Seoul, Pyongyang, Tokyo, and Peking all have a voice.

One promising avenue for promoting such cooperative efforts might lie in the creation of representative regional machinery designed to bring together technical experts on a nonpolitical basis, possibly but not necessarily under the auspices of the United Nations. A modest but promising start in this direction has been made in the form of the Committee for the Coordination of Joint Prospecting for Mineral Resources in Asian Offshore Areas (CCOP), a grouping of nine

East Asian and Southeast Asian countries established in 1966 at the instance of Taiwan, South Korea, and Japan as an agency of the United Nations-sponsored Economic Committee for Asia and the Far East (ECAFE).[13] However, the CCOP would have to be broadened in its membership or, more probably, superseded by new machinery, since it is deeply suspect in the eyes of Peking, Hanoi, and Pyongyang as a result of its initial sponsorship and its extensive offshore studies in disputed areas where Taipei, Seoul, and Tokyo have staked out concessions.

The CCOP has been notably unsuccessful in recent efforts to enlist the participation of Peking, if only because it is regarded as a creature of American influence. This is supremely ironic, since the CCOP has received relatively little support from the U.S. government and has been treated with studied reserve by Western oil companies, who see a threat to their interests in the very idea of an intergovernmental agency conducting public oil survey operations. Nevertheless, its origins did lie in the cold war desire of U.S.-supported regimes to preempt offshore areas that might otherwise be developed by China. It would be extremely difficult to enlist Peking in a new regional effort without setting up a different organization or, at the very least, reconstituting the CCOP on an entirely new basis, perhaps along subregional lines, with the recently formed ASEAN Council on Petroleum (ASCOPE) as a Southeast Asian arm and another branch in northeast Asia embracing Peking. The most propitious time for such an initiative would be after the status of Taiwan has been resolved, when the issue of membership for Taipei would no longer be divisive. Hanoi would occupy the place once held by Saigon as the Vietnamese member in such a reorganization, but Seoul and Pyongyang would both be initially excluded from formal membership as disputants in an unresolved civil conflict. Hopefully, some of the technical subcommittees of such a body could provide a platform where Seoul and Pyongyang might eventually be induced to participate jointly in ad hoc studies relating to geologically interlinked areas between Korea and China in the Yellow Sea; between Korea and Japan in the Sea of Japan; and between Korea, China, and Japan in the East China Sea. The magnitude of the economic stakes involved could make such joint participation much more attractive to both parties than other proposals for North-South cooperation hitherto advanced.[14] In particular, the geological potential of the East China Sea sector embraced in the 1974 Seoul-Tokyo agreement (Figure 7) is regarded as extremely encouraging by the companies so far involved there. A desirable long-term objective would be some form of unified North-South representation in Sino-Japanese-Korean negotiations addressed to the delimitation of boundary rights and the planning of coordinated development arrangements in the East China Sea in which all parties would share the costs and benefits. In some respects, the 1974 agreement could serve as a model for a broadened agreement established on more enduring political foundations.[15] But progress in this direction would presuppose an overall improvement in the relations between North and South as well as between China and Japan.

Despite the many explosive aspects of the offshore disputes emphasized in this book, there are several substantial factors that could work to promote peaceful boundary settlements and possibly even cooperative offshore activity. The most important of these is the high cost and speculative character of deep-water exploration and development. As their offshore experience grows, Peking and its Asian neighbors alike will become increasingly aware of the enormous financial risks involved and thus more disposed over the years to consider collaborative ventures and revenue-sharing in disputed areas as a means of diffusing these risks on a bilateral or multilateral basis. This would be particularly true with respect to any discoveries of natural gas, since it is so much more expensive to develop offshore gas than offshore oil. By the same token, in the absence of a cooperative approach these financial considerations could paralyze offshore development indefinitely, reinforcing the political constraints resulting from unresolved boundary conflicts.

Given its rivalry with Moscow, Peking might well feel compelled to adopt a cooperative posture for strategic reasons, even in cases where the potential economic rewards of unilateral action might seem most tempting. An aggressive Chinese posture could open up opportunities for Soviet political intervention in East Asia by aggravating tensions with neighboring countries. By contrast, a conciliatory approach could be utilized to strengthen Peking's already preponderant position in regional power relationships and thereby to limit Soviet incursions. Suppose, for example, that China were to offer oil rights in the Senkaku (Tiao-yü T'ai) area to Japan in return for formal acknowledgment of Chinese sovereignty over the islands and payment of a stipulated share of any resulting revenues. What Peking would lose economically in such an arrangement would be more than offset politically. For all practical purposes, Japan would be confirming China's position of regional primacy, and as its investment outlays multiplied, Tokyo would feel increasingly beholden to Peking. Similarly, in Korea, as suggested in chapter 5, a cooperative approach on oil issues could be employed to encourage a softened Southern posture toward both Pyongyang and Peking.

The competition for offshore resources in East Asia will ultimately be shaped by the hard reality that China is steadily adding to its power advantage over its neighbors and will be able to employ both carrot and stick in promoting its own concept of cooperation. Already dominant militarily, Peking is gradually enhancing its regional military superiority, notably in the critical arena of naval power most directly relevant to potential conflicts over offshore resources. With each passing year, Peking's oil-hungry Asian neighbors will face increasingly difficult choices. On the one hand, the compromises necessary to facilitate oil development could prove extremely painful, possibly requiring the recognition of

Chinese sovereignty in cases such as the Senkakus (Tiao-yü T'ai) and some of the South China Sea islands. On the other, more determined bargaining by the countries concerned could mean a continued stalemate on boundary issues and the perpetuation of their overwhelming dependence on the Middle East and the Western majors. Over time, such a determined stance could also provoke unilateral action by Peking in pursuit of maximum Chinese claims. The search for a way out of this dilemma during the decades ahead will not only provide a revealing barometer of the shifting geopolitical climate in East Asia itself. Even more important, it will indicate whether the region is emerging as a coordinated power center under Chinese leadership in the larger arena of world politics.

Notes

1. The *Gulfrex* Decision

1. Like most of this book, the account of this meeting and the policymaking process of which it was a part is based primarily on interviews with many of the principals concerned. In this case, the author has also consulted State Department and Defense Department documents made available in response to inquiries under the Freedom of Information Act. Out of 137 relevant documents located in its files, the State Department released 17, with substantial deletions in several (Case No. 5-D-334; State Department policy is set forth in a letter from Lester E. Edmond, deputy assistant secretary for East Asian and Pacific affairs, 8 October 1975). Five additional documents were made available, with deletions, by the Defense Department (Case No. DF01-981; Defense Department policy is set forth in a letter from Charles W. Hinkle, director, Freedom of Information and Security Review, 17 December 1975).

2. A series of thirteen messages was sent seeking instructions, cul-

minating in two on 30 December. In a message on 14 December (5343/140950), Ambassador Walter McConaughy warned that "anyone boarding with a pistol" could take over the *Gulfrex* or Oceanic Exploration's vessel, *Western Beach*. In a message on 21 December (210940Z), CINCPAC spelled out Chinese naval capabilities in the area concerned, pointed out that the existing Rules of Engagement "do not provide for the reaction of U.S. military forces to hostile acts against non-military contract U.S. ships," and concluded that the proximity of Taiwan's concession zones to the Chinese mainland, coupled with explicit Chinese warnings against oil exploration, "leads to the conclusion that possible harassment of the survey ships, or more severe action by Chinese patrol craft, cannot be discounted." The *Gulfrex* was already in regular contact with U.S. ships in the area, CINCPAC said, and daily position reports should be made by other survey vessels operating thenceforth in the area to facilitate rapid military action in their defense.

3. K. O. Emery et al., "Geological Structure and Some Water Characteristics of the East China Sea and the Yellow Sea," *CCOP Technical Bulletin*, United Nations ECAFE (Bangkok, May 1969), 2:41.

4. Letter: Packard to Harrison, 22 October 1975.

5. SECDEF No. 9201/011727Z, 1 January 1971 (White House/State/Defense/JCS message). The text of this message was not released.

6. Memorandum to Herman Barger, deputy assistant secretary of state, 6 January 1971 (1-196/71). Other salient declassified documents relating to the creation of the task force are a Memorandum for Record, Joint Chiefs of Staff, Plans and Policy Directorate, signed by Brigadier General F. L. Smith, USAF, chief, Far East Division, J-F, 13 January 1971, and a memorandum from Acting Assistant Secretary of State for Far Eastern Affairs Winthrop Brown and legal adviser John Stevenson to Secretary of State Rogers, 12 January 1971.

7. On the oil issue, State Department spokesman Charles Bray said that China "has asserted a claim to large and imprecisely defined areas of the continental shelf" and that the U.S. government had "informed the companies that under the circumstances we consider it inadvisable for them to undertake operations in these disputed areas." (Murray Marder, "U.S. Cautions Oil Seekers near China,"

Washington Post, 10 April 1971, p. 1. For a Japanese perspective on the announcement, see "Senkakus Sovereignty Issue Should Be Solved in Talks," *Japan Times,* 11 April 1971, p. 1).

8. North and South Korea, Japan, Taiwan, Malaysia, Vietnam, Indonesia, Brunei, and the Philippines.

2. China's Oil Potential: Problems and Prospects

1. Tatsu Kambara, "Chugoku no Sekiyu Sangyo" [The Chinese petroleum industry], *Sekiyu Kaihatsu Jiho* [Oil development reports] (Tokyo, December 1974), 24:37, Table VII. An abridged version of this study, "The Petroleum Industry in China," appeared in *China Quarterly* (October/December 1974), pp. 699–719.

The 1974 estimate appears in *China: Energy Balance Projections,* Central Intelligence Agency (November 1975), p. 33, Table IX; see also p. 21.

All energy-demand figures cited in this chapter are expressed in terms of oil equivalence.

2. Masanobu Otsuka, "Chugoku no Energi Keizai to Taigai Boeki" [China's energy economy and foreign trade], *Chugoku Keizai Kenku Geppo,* Japan External Trade Organization, 15 April 1975, especially pp. 30–36, 51–56, 69–77, 79–83. See also "Chugoku Keizai no Hatten to Energi Jijo" [Chinese economic development and the energy situation], *Zaikai Kansoku* [Financial survey] (Tokyo, April 1975), pp. 1–31.

3. Thomas G. Rawski, "The Role of China in the World Energy Situation" (background paper prepared for a Brookings Institution study, June 1973). The findings in this paper are summarized by Thomas G. Rawski, "China and Japan in the World Energy Economy," in E. W. Erickson and Leonard Waverman, eds., *The Energy Question: An International Failure of Policy* (2 vols.; Toronto: University of Toronto Press, 1974), 1:105–7.

Otsuka assumes a less rapid rate of agricultural mechanization and household energy use than Rawski, but both make broadly similar assumptions with respect to the growth of energy demand in transportation and the unit of energy consumption per dollar of added value.

4. Masahiko Ebashi, "Outlook on China's Foreign Trade and Oil Exports," *JETRO China Newsletter* (Tokyo, April 1975), especially p. 23, Table 2.

5. In "Petroleum Industry in China," Kambara assumes a 36 percent share for oil.

6. Hideo Ono cites this production target in "China's Crude Oil and Price Supply Outlook," *JETRO China Newsletter* (Tokyo, January 1975), p. 13. Ono states that China's "second long-term plan" for oil development began in 1970 and included interim targets of 168.4 million tons (1.3 billion barrels) by 1980 and 309 million tons (2.3 billion barrels) by 1985. This has not been authenticated elsewhere in detail. However, numerous Japanese visitors to the Taching oil field have been told that a goal of one ton of oil for every two Chinese, or 400 million tons (3 billion barrels), had been set as a national target for 1990 by the late "Iron Man," Wang Chen-hsi, a venerated pioneer of the Chinese oil industry, shortly before his death in 1970. For example, see Otsuka, "Chugoku no Energi Keizai to Taigai Boeki," p. 31; *Mainichi* (Tokyo), 7 July 1974, p. 1; and an article by Ryuzo Yamashita in *Ekonomisuto* (Tokyo), 3 September 1974.

7. For useful summaries of the background of Chinese oil development and the progress of existing fields, see Bobby A. Williams, "The Chinese Petroleum Industry: Growth and Prospects," *China: A Reassessment of the Economy*, Joint Economic Committee, U.S. Congress (July 1975), pp. 225–63; and C. Y. Cheng, *China's Petroleum Industry: Output Growth and Export Potential* (New York: Praeger, 1976).

8. Oil production increased at an average rate of 24.6 percent from 1963 through 1974. In 1975 it increased by an estimated 16 percent over 1974. However, statistics for the first quarter of 1976 indicated a drop to a 12.7 percent increase over the comparable period of 1975.

9. The 1971 and 1974 figures are discussed in Williams, "Chinese Petroleum Industry," p. 241. The 1975 figure is cited by the New China News Agency in a 12 November 1975 broadcast reported in the *U.S. Foreign Broadcast Information Service Daily Report*, National Technical Information Service, Department of Commerce, 13 November 1975, p. E2. *Current Scene*, a publication of the U.S. Consul-

ate in Hong Kong, estimated an 11 to 12 percent growth rate in coal for 1975 in its June 1976 issue. However, this estimate was derived in a manner questioned by many experts; reported growth rates in coal published by some provinces were related to previous production estimates for those provinces and then given a weightage consistent with the past share of national output produced by the province in question. On this basis, production levels were assumed for provinces that did not publish growth rates.

10. *China: Energy Balance Projections,* p. 8.

11. New China News Agency broadcast, 1 November 1975, reported in the *U.S. Foreign Broadcast Information Service Daily Report,* National Technical Information Service, Department of Commerce, 7 November 1975, p. E2.

12. V. I. Akimov, "The Fuel and Power Base of the People's Republic of China," *Problems of the Far East,* no. 1, Joint Publications Research Service, no. 61955, 9 May 1974, p. 68, a translation from *Problemy Dal'nego Vostoka* [Problems of the Far East], no. 1 (Moscow, 1974).

13. Informed Japanese sources estimate that China's absorptive capacity in refineries and petrochemical plants had reached 59–61 million tons in 1974, although Soviet observers in particular question this figure. In any case, the 65-million-ton production estimate for 1974 accepted by most U.S. and Japanese observers would leave a gap of 5–10 million tons, depending on the rate assumed for the utilization of refinery capacity and the amount stored for military and other purposes.

14. Thermal efficiency is the output of mechanical work per unit of raw energy input. For the best discussion with reference to China, see Rawski, "Role of China in World Energy Situation," pp. 8*ff.*

15. The high wax content of these grades makes it necessary to install costly "cracking" facilities to refine the oil, and even then the yield of gasoline and other distillates is relatively low unless the crude is blended with other grades. See chapter 7 for a discussion of this problem in relation to Chinese oil exports.

16. *China: Energy Balance Projections,* p. 8.

17. For the most complete discussion of China's petrochemical industry in English, see Sy Yuan, "China's Chemicals," *U.S. China*

Business Review (November/December 1975), pp. 37–53. For the best description in English of the Liao-yang complex, see "Mainland Chinese Aim High," *Oil and Gas Journal,* 10 November 1975, pp. 197–203.

18. Keiji Samejima, "Chugoku Netsu Komoru Kaitei Yuden Kaihatsu" [A new period of activity for China's petrochemical industry], *Nihon Keizai* (Tokyo), 14 February 1975, p. 7.

19. Ebashi, "Outlook on China's Foreign Trade," p. 16, and p. 25, Table 5.

20. Williams, "Chinese Petroleum Industry," pp. 247–49.

21. *China: Energy Balance Projections,* especially pp. 13, 29.

22. Cheng, *China's Petroleum Industry,* p. 40, Table 2.13.

23. *China: Energy Balance Projections,* p. 19. The CIA study cites a Soviet growth rate in petroleum production of 6.5 percent for the 1965–73 period. Further downgrading Chinese production prospects, a 1977 CIA study said that "the reserve and production outlook is much less favorable than it appeared a few years ago." *The International Energy Situation: Outlook to 1985* (April 1977), p. 13. An average Soviet rate of 11.3 percent for the 1954–72 period is cited by Vaclav Smil in "Communist China's Oil Exports: A Critical Evaluation," *Issues and Studies* (Taipei, March 1975), p. 75. Reviewing the 1955–65 period, Robert E. Ebel found a 13.8 percent Soviet rate. See Ebel, *Communist Trade in Oil and Gas* (New York: Praeger, 1970), p. 40.

24. Cheng, in *China's Petroleum Industry,* does not go beyond 1985, but even 10 percent growth after 1985 would result in 405.3 million tons (3 billion barrels) in 1987. His study was adapted from an earlier analysis for the Department of Commerce, *The U.S. Export Potential of Petroleum Equipment to the People's Republic of China,* Contract No. 4-36289, August 1974. Meyerhoff predicts 332 million tons by 1985 in "Petroleum Geology and Industry of the People's Republic of China," *CCOP Technical Bulletin,* United Nations ESCAP (Bangkok, July 1977).

25. Saudi Arabia has deliberately kept its production below its capabilities. In 1974 facilities already available would have permitted a level of production at least 30 percent higher than 412 million tons (3.1 billion barrels).

26. Ebashi, "Outlook on China's Foreign Trade," p. 26, Table 6.

27. *China: Energy Balance Projections,* p. 13.

28. For a discussion of the trade-offs between short-term and long-term growth associated with oil development, see K. C. Yeh and Y. L. Wu, "Oil and Strategy" (paper prepared for the Conference on China sponsored by the Institute of International Relations, Taipei, May 1976), especially pp. 21–24.

29. Cheng, *China's Petroleum Industry,* pp. 189–90.

30. Randall W. Hardy, "China's Oil Potential," mimeographed (Federal Energy Agency, June 1976), pp. 35–37.

31. Cheng, *China's Petroleum Industry,* pp. 109–11.

32. For example, see "Wo Kuo Hai-yang Ti-chih Tiao-Ch'a Shih-ye Hsün-su Fa-chan" [Our work of exploring the ocean's geology has developed rapidly], *Kuangming Jih-pao,* 2 May 1976, p. 1.

33. For a useful review of the history of Chinese efforts to develop indigenous manufacturing facilities and an enumeration of the thirty-one major centers, see Cheng, *China's Petroleum Industry,* pp. 108–24.

34. Alice Tisdale Hobart, *Oil for the Lamps of China* (New York: Grosset and Dunlap, 1933).

35. See Gustav Egloff, "China's Potential Oil Resources Large," *Oil and Gas Journal,* 28 December 1946, pp. 243–48; and Charles A. Heller, "Oil for the Lamps of China—and Russia?" *World Petroleum* (1963), pp. 58–78.

36. A. A. Meyerhoff, "Developments in Mainland China, 1949–1968," *American Association of Petroleum Geologists Bulletin* (August 1970), pp. 1575–78. Meyerhoff refers to his access to Soviet data in "Geopolitical Implications of Russian and Chinese Petroleum," in Virginia S. Cameron, ed., *New Ideas, New Methods, New Developments,* vol. 2, *Exploration and Economics of the Petroleum Industry* (New York: Matthew Bender, 1974), p. 91. This publication is a report of the Institute on Petroleum Exploration and Economics, 14–15 March 1973.

37. Interviews by the author. The figure of 10 billion tons (75 billion barrels) is also cited by Paul H. Fan, professor of geology, University of Houston, on the basis of letters from former schoolmates in China. (Paul H. Fan, "Chinese Oil Industry Image Changing,"

Oil and Gas Journal, 11 August 1975, p. 112; see also a reference to the 10-billion-ton figure in *Asia Research Bulletin*, 31 January 1976, p. 163.)

38. "World Crude Resources May Exceed 1,500 Billion Barrels," *World Oil* (September 1975), p. 48, Table 2. This is based on a widely cited paper presented by J. D. Moody at the Ninth World Petroleum Congress in Tokyo, May 1975.

39. See John D. Hawn, ed., *Methods of Estimating the Volume of Undiscovered Oil and Gas Resources* (Tulsa: American Association of Petroleum Geologists, 1975).

40. A. A. Meyerhoff, "China's Petroleum Potential," *World Petroleum Report, 1975*, p. 21.

41. Williams, "Chinese Petroleum Industry," pp. 225, 234–35.

42. Bobby A. Williams, "The Petroleum Industry in China: A Note" (unpublished paper, March 1976), p. 11.

43. "Oil Deposits in China," *JETRO China Newsletter* (Tokyo, January 1975), p. 24.

44. N. J. Sander (AMOCO International Oil Company), John F. Mason (Continental Oil Company), and William E. Humphrey (AMOCO International Oil Company), "Tectonic Framework of Southeast Asia and Australasia: Its Significance in the Occurrence of Petroleum" (paper presented at the World Petroleum Congress, Tokyo, May 1975), see especially Tables 1–3.

45. Hardy, "China's Oil Potential," Appendix III.

46. Interview by the author with Mobil Oil Company geologists.

47. Hardy, "China's Oil Potential," Appendix III.

48. Williams, "Petroleum Industry in China: A Note," p. 9.

49. J. S. Lee, *The Geology of China* (London: Thomas Murphy, 1939), p. 319.

50. For a useful discussion of plate tectonics in relation to China, see Frank Press, "Plate Tectonics and Earthquake Prediction: Contrasting Approaches in China and the United States," *Bulletin of the American Academy of Arts and Sciences* (May 1975).

51. For example, see Hideyuki Matsuishi, "Chugoku no Engan Kaiyo Kaihatsu" [China's coastal oceanographic exploration],

Keidanren Geppo, (Tokyo, April 1975), pp. 43–47. This is an account of a mission of Japanese oceanographers to China sponsored by Keidanren, Japan's leading business federation.

52. Yen Tun-shih, "Outline of the Meso-Cenozoic Tectonic Framework of Eastern China" (paper presented at the Lamont-Doherty Geological Observatory, Palisades, N.Y., October 1975), pp. 2–7. Portions of this paper are paraphrased in Walter Sullivan, "Scientists Say Collisions of Three Continents Formed Asia," *New York Times,* 9 October 1975, pp. 1, 48.

53. Lee, *Geology of China,* especially pp. 247, 326–39.

54. Yen, "Tectonic Framework of Eastern China," pp. 9–11.

55. See Maurice J. Terman, "Cenozoic Tectonics of East Asia" (paper presented at the Thirteenth Pacific Science Congress, Vancouver, September 1975).

56. Sullivan, "Collisions of Continents Formed Asia."

57. For a description of the Uinta Basin, see R. T. Ryder et al., "Early Tertiary Sedimentation in the Western Uinta Basin, Utah," *Geological Society of America Bulletin* (April 1976), pp. 496–512.

58. Terman estimates that each "graben" structure in northeast China is likely to contain 133 million tons (1 billion barrels) of oil. Mungan, research director at the Petroleum Recovery Institute, University of Calgary, Alberta, sees 400 or 533 million (3 or 4 billion) as more likely and considers 800 million (6 billion) "distinctly possible."

59. For example, see "Shantung K'uai-tuan Kuo-tsao T'e-ch'ang Yu Ti-chen-tai Te Ts'u-pu Ch'u-fen" [Characteristics of the block-faulting tectonics of the Shantung region], *Report of the Institute of Geology,* Academia Sinica (Peking, 1974), English abstract, p. 329.

60. *Peking Review,* 24 May 1974.

61. See Hardy, "China's Oil Potential," p. 7 and Appendix IV. The 2.7-billion-ton (20.3-billion-barrel) base for eastern China came out of an overall onshore reserve estimate of 10.1 billion tons (75.5 billion barrels).

62. Following a visit to Taching in 1976, Necmettin Mungan said in an interview and in a letter to me (4 August 1976) that the Chinese had shifted from the "direct line drive" method to the "inverted

nine spot" method. In the former, there is one production well for each water injection well, which results in wastage of oil, while in the latter, three production wells share the amount of water otherwise injected into one well. Mungan said that the change could double the amount of oil obtained but would result in a decline in the rate of production in each area drilled. However, he has explained, "the Chinese seem to believe that they will find enormous quantities of oil" and that the conversion to the new methods will not necessarily affect their overall national rate of production. (See "Draft Report of a Seminar on China's Energy Policies and Development" (Stanford University, 2–3 June 1976), p. 38. Most Chinese oil fields also appear to have abandoned the "direct line drive" method, though information is incomplete.

63. Kim Woodard, "The International Energy Policies of the People's Republic of China" (Ph.D. dissertation, Stanford University, 1976), vol. 2, "Statistical Profile," Table XA2 ("Low Growth Scenario"), p. 646, and Table XC1 ("High Growth Scenario"), p. 650. Woodward presents a grand total of 432 alternate projections in his discussion of China's future energy balance, utilizing four different reserve scenarios, four reserves conservation policies, three energy–GNP coefficients, three per capita GNP growth-rate scenarios, and three population-growth scenarios.

3. Another Persian Gulf?

1. For example, see Monty Hoyt, "China Seas Oil Bonanza," *Christian Science Monitor*, 15 October 1971, p. 1.

2. See Hideyuki Matsuishi, "Chugoku no Engan Kaiyo Kaihatsu" [China's coastal oceanographic exploration], *Keidanren Geppo* (April 1975), 23(4):45. Matsuishi reported that "the existence of large scale seabed oil fields comparable to those in the Persian Gulf has been confirmed in the Gulf of Pohai." Similarly, Ho Ping-ti, a University of Chicago historian and president of the Association of Asian Studies in 1975, wrote after a visit to China in 1974 that "authoritative sources in Peking" had told him that the size of China's known oil deposits, not counting the "unexplored and potentially rich" areas of the Yellow Sea and the East China Sea, are "larger than the presently known reserves of the entire Middle East." Specifically, Ho declared that "the Pohai Gulf is destined to be the second Per-

sian Gulf of the world." Ho Ping-ti, "China Is the Richest Country in Oil Reserves—A New and Important Factor for Chinese Economic Development in the Seventies," *The Seventies* (Hong Kong, February 1975), pp. 7–9. See also *Los Angeles Times,* 13 October 1974, Part 6, p. 1.

3. A. A. Meyerhoff, "China's Petroleum Potential," *World Petroleum Report 1975,* p. 21.

4. In *Sotsialisticheskaya Industrya* [Socialist industry], 1 August 1974, F. Salmanov declares that the East China Sea may "possibly" contain reserves of 1 to 1.5 billion tons (7.5 to 11.2 billion barrels), discounting reports that the Chinese offshore areas contained "hundreds of billions of tons." This is cited in the *Review of Sino-Soviet Oil* (September 1974) and the *Oil and Gas Journal,* 7 October 1974, p. 53.

5. Jan-Olaf Willums, "China's Offshore Oil: Application of a Framework for Evaluating Oil and Gas Potentials under Uncertainty" (Ph.D. dissertation, Massachusetts Institute of Technology, 1975), especially p. 38.

6. Estimates made in interviews tended to fall within 2.6 and 6 billion tons (20 and 45 billion barrels). In his Federal Energy Agency study, "China's Oil Potential," Randall W. Hardy cites estimates of 2.7 billion and 1.3 billion tons (20 billion and 10 billion barrels) by two different majors (Appendix III). These constitute 21 percent and 25 percent of the total reserve estimates for China made by the two companies concerned.

7. Jan-Olaf Willums, "Prospects for Offshore Oil and Gas Developments in the People's Republic of China" (paper no. 2086, Offshore Technology Conference, Houston, May 1974), p. 542. So far as I can determine, this estimate applies to the area covered by the provisional Japanese concessions held by the Uruma Oil Company (Figure 6) and the semigovernmental Sekiyu Kaihatsu Company.

8. In evaluating this estimate, allowance should be made for the fact that Clinton has been seeking to enlist collaborators in developing this concession.

9. "The Wealth of the Oceans," *Ko-Hsüeh Shih-yen* [Scientific experiment], no. 4 (1973). See also "China Tops 1973 Petroleum Production Plan," *Economic Report* (English supplement) (Hong Kong, January/March 1974), p. 6.

10. New China News Agency broadcast in English at 0724 (GMT), 18 September 1974, and reported by Reuters (Hong Kong).

11. "China Will Become a Main Oil Producing Country at the End of the 1970s," *Tien Tien* (Hong Kong), 6 August 1974, p. 1. Similarly glowing accounts of reserve prospects may also be found in *Ta-kung Pao* (Hong Kong), 19 September 1974, p. 1, and in a Peking-datelined account in *Shangpao* (Hong Kong), 20 September 1974, p. 1.

12. Reported in *Hsingtao* (Hong Kong), 25 December 1974, p. 1. Several references to "bright" reserve prospects may be found in Hua Ching Yuan, "New Achievements in China's Oil Industry," *China's Foreign Trade*, no. 1 (1975), p. 6.

13. Hiroshi Niino and K. O. Emery, "Sediments of Shallow Portions of the East China Sea and the South China Sea," *Bulletin of the Geological Society of America* (1961), 72:731–62.

14. "Stratigraphy and Petroleum Prospects of the Korea Strait and the East China Sea," *Report of Geophysical Exploration*, vol. 1, no. 1 (1967), reprinted in the *Geological Survey of Korea* (June 1971) and in the *CCOP Technical Bulletin*, vol. 1, United Nations ECAFE (Bangkok, June 1968).

15. M. V. Klenova, "Sediment Maps," *Oceanologia et Limnologia Sinica* (Peking, 1958), 1(2):243–54.

16. *CCOP Technical Bulletin* 1:17, 27.

17. *Ibid.*, p. 13. Emery's 1969 report, cited in footnote 18, also noted that "in general, the fill is thickest, widest and most continuous in the Okinawa Trough" (p. 39).

18. *Ibid.*, p. 25.

19. K. O. Emery et al., "Geological Structure and Some Water Characteristics of the East China Sea and the Yellow Sea," *CCOP Technical Bulletin*, United Nations ECAFE (Bangkok, May 1969), 2:41.

20. *Ibid.*, p. 4.

21. Emery et al., "Geological Structure and Some Water Characteristics," pp. 31, 35.

22. In "Cenozoic Tectonics of East Asia" (paper presented at the Thirteenth Pacific Science Congress, Vancouver, September 1975), Terman sketches his concept of the "North China–Korean plate,"

NOTES: 3. ANOTHER PERSIAN GULF? 277

showing its continuation from the Shantung Peninsula into adjacent offshore areas. Terman, who reads Chinese, has based his analysis on Chinese seismological publications as well as on the findings of the Earth Resources Technology Satellite, an international research venture in plate tectonics, and on tectonic mapping studies for the Advanced Research Project Agency of the Defense Department. In itself, the Earth Resources Technology Satellite does not add to the geological knowledge of the offshore areas. However, an ambitious project to build a computerized map of the mineral resources of all areas bordering the Pacific Ocean was undertaken at the Circum-Pacific Energy and Minerals Resources Conference held in Honolulu in August 1964. (See the report on this conference prepared by the Subcommittee on Mines and Mining of the Committee on Interior and Insular Affairs, U.S. House of Representatives, October 1974, pp. 5–8.) It was in connection with this conference that Terman and other U.S. scientists have begun to marshal data that complement the studies under way in China itself with respect to the tectonic factors shaping Chinese petroleum geology.

23. S. B. Frazier et al., "Marine Petroleum Exploration of the Huksan Platform: Korea" (paper presented at the Circum-Pacific Energy and Minerals Resources Conference, Honolulu, August 1974), especially p. 3 and Fig. 2.

24. N. J. Sander, John F. Mason, and William E. Humphrey, "Tectonic Framework of Southeast Asia and Australasia: Its Significance in the Occurrence of Petroleum" (paper presented at the 1975 World Petroleum Congress, Tokyo, May 1975), p. 20.

25. *Ibid.*, p. 12.

26. Maurice Mainguy, "Regional Geology and Petroleum Prospects of the Marine Shelves of Eastern Asia," *CCOP Technical Bulletin*, United Nations ECAFE (Bangkok, May 1970), 3:103–5.

27. Yasufumi Ishiwada, "Nippon Retto Shuhen Tairiku Dankyo no Sekiyu Kishitsu" [Petroleum geology of the continental shelf surrounding the Japanese islands], *Sekiyu Gakkai-shi* [Oil association journal] (1975), 18(6):27.

28. Mainguy, "Regional Geology and Petroleum Prospects," p. 105.

29. *Metallogenesis, Hydrocarbons and Tectonic Patterns in East Asia*, United Nations Development Programme (CCOP) (Bangkok, 1974),

p. 56. This publication is the report of the International Decade of Ocean Exploration Workshop held in Bangkok 24–29 September 1973.

30. K. O. Emery and Zvi ben-Avraham, "Structure and Stratigraphy of the China Basin," *CCOP Technical Bulletin*, United Nations ECAFE (Bangkok, July 1972), 6:117, 138.

31. M. L. Parke, Jr. et al., "Structural Framework of the Continental Margin in the South China Sea," *CCOP Technical Bulletin*, United Nations ECAFE (Bangkok, June 1971), 4:140.

32. "Regional Geology and Offshore Prospects for Minerals in the Republic of Vietnam," *Report of the Fourth Session of the CCOP*, United Nations ECAFE (E/CN.11/L.190), 31 January 1968, p. 70.

33. Willums, "China's Offshore Oil," especially Table 8-2 and pp. 38 and 274–81. Now manager of special projects for the Saga Petroleum Company, Oslo, Norway, Willums developed his estimates by utilizing the Bayesian Probability Theory, a method suggested by MIT mathematician Gordon M. Kaufman in his work *Statistical Decisions and Related Techniques in Oil and Gas Exploration* (Englewood Cliffs, N.J.: Prentice-Hall, 1969). Willums attributes his geological methodology and data primarily to Douglas Klemme, vice-president for exploration of Weeks Natural Resources.

4. China Goes Offshore

1. Ch'in Yun-shan and Hsu Shan-min, "Bottom Sediments of the Northern Yellow Sea and Po Hai" (unpublished draft, 1958); and Ch'in Yun-shan and Fan Shih-ch'ing, "Preliminary Study of Submarine Geology of China's East Sea and the Southern Yellow Sea," *Translations on Communist China*, no. 97, Joint Publication Research Service 50252, 7 April 1970, pp. 12–36, translation from the original in *Hai-yang Yü Hu-chao* [Oceans and lakes] (1959), 2:82–84. See also H. K. Wong and T. L. Ku, "Oceanography and Limnology in Mainland China," *CCOP Technical Bulletin*, United Nations ECAFE (Bangkok, May 1970), 3:137–46.

2. Chang Chi-wu, "Getting Minerals from the Ocean Bottom," *Translations on Communist China*, no. 6, Joint Publication Research Service 44937, pp. 40–42, translated from the original in *K'o-Hsüeh*

Hua-pao [Science illustrated] (Shanghai, April 1960), p. 142. See also Wen Ts'ao, "Penetrating and Exploiting the Oceans," *K'o-Hsüeh Hua-pao* [Science illustrated] (Shanghai, April 1960), pp. 138–39, in *Translations on Communist China, no. 6.*

3. Ch'in Yun-shan and Cheng T'ieh-min, "Preliminary Study of Bottom Sediments of the Coastal Sea of Chekiang" (unpublished draft, 1961), cited in Ch'in Yun-shan, "Initial Study of the Relief and Bottom Sediment of the Continental Shelf of the East China Sea," *Hai-yang Yü Hu-chao* [Oceans and lakes] (1963), 5:35. See also Ch'in Yun-shan and Liao Hsien-kuei, "Preliminary Discussion of Sedimentation Function of the Po Hai Gulf," *Hai-yang Yü Hu-chao* [Oceans and lakes] (1962), 4:199–207.

4. Hiroshi Niino and K. O. Emery, "Sediments of Shallow Portions of the East China Sea and the South China Sea," *Bulletin of the Geological Society of America* (1961).

5. Ch'in, "Initial Study of Relief and Bottom Sediment," especially pp. 18–19, 24, 33–34.

6. For a discussion of the Mattei deal with China, see "Italian Oil Chief Dies in Air Crash," *New York Times,* 28 October 1962, p. 16; and C. L. Sulzberger's column, "Foreign Affairs," *New York Times,* 20 August 1962, p. 22. See also Charles R. Dechert, *ENI, Profile of a State Corporation* (Leyden: E. J. Brill, 1963), p. 47.

7. Takahashi Shogoro, "Bokkai no Kaiyo Sekiyu Kaihatsu o Megutte" [Concerning the development of offshore oil in the Po Hai Gulf area], *Kaiyo Sangyo Kenkyu Shiryo* [Sources on offshore industrial research], Institute for Ocean Economics (Tokyo), 20 November 1973, p. 2.

8. *Ibid.*

9. "Bokkai (Daiko) Kaitei Yuden no Kaihatsu Katei" [Offshore oil field development in the Po Hai Gulf (Takang)]. This study appeared in the house organ of a Japanese oil company in 1973 and was made available by an American oil-equipment company on the understanding that the identity of the Japanese company would not be cited.

10. The deal reportedly involved two U.S. companies, Gulf and Union Oil; the Japanese trading company C. Itoh, acting as middleman; and an unidentified rig company. Union Oil owns 20 percent

of the Japanese oil company Maruzen, which operates refineries in Shimotsu, Matsuyana, and Chiba. These refineries utilize Kuwait crude imported to Japan by Gulf. Gulf was anxious to demonstrate its good will to China in the wake of the Kissinger visit with an eye to possible future purchases of Chinese oil for its refineries in Japan and helped to arrange for the manufacture of the rig by a Japanese firm under a licensing arrangement.

11. *U.S. China Business Review* (January/February 1974), p. 31. Asia Offshore Drilling is a subsidiary of Teikoku Oil, which is affiliated, in turn, with Gulf, mentioned earlier as the prime mover instrumental in arranging the sale of the jack-up.

12. Hideyuki Matsuishi, "Chugoku no Engan Kaiyo Kaihatsu" [China's coastal oceanographic exploration], *Keidanren Geppo* (April 1975), 23(4):43.

13. "China's Oil Rich Continental Shelf Eyed for Exploitation," *Wen-hui Pao*, 18 March 1973, p. 3.

14. "Mainland China," *Offshore*, 20 June 1975, p. 227.

15. Hsiao Hsi-shu and Tseng Hen-i, "The *Pohai No. 1* Offshore Drilling Rig," *K'o-Hsüeh Shih-yen* [Scientific experiment], no. 1 (January 1975), p. 3. The author has consulted an internal State Department translation.

16. "China's First Oil-drilling Ship," *China Reconstructs* (Peking, July 1976), pp. 36–37.

17. "Another China-made Research Vessel—*Pohai I*—Sails Out for Drilling Test," *Ta-kung Pao*, 13 January 1975, p. 1.

18. Hsiao and Tseng, "*Pohai No. 1* Offshore Drilling Rig."

19. "China's First Oil Drilling Ship," pp. 37, 40.

20. Letter: Sage to Harrison, 20 August 1975. Sage explained that the total length of the legs is not the effective length because part of the legs reach into the hull of the rig, an "air gap" is needed to allow for wave action, and an additional portion of the total leg length is needed for penetration into the ocean floor.

21. Known in the rig industry as the ETA design.

22. "China-made Deep-sea Rig Successfully Drills an Oil Well in Yellow Sea," *Ta-kung Pao*, 31 December 1974, p. 1.

23. "Wo Kuo Hai-yang Ti-chih Tiao-Ch'a Shih-ye Hsün-su Fa-

chan" [Our work of exploring the ocean's geology has developed rapidly], *Kuangming Jih-pao*, 2 May 1976, p. 1.

24. The photograph was published in *China Pictorial* (January 1975), p. 13. The rig was likened to the M. W. Thornton model operated by the Reading and Bates Drilling Company.

25. See *Asahi*, 22 June 1973, p. 1; *Sekiyu Kaihatsu* [Oil development], vol 10, no. 124 (1973); and *China's Oil Industry: A Background Survey* (working paper presented by the Sino-British Trade Council, London, to a conference on "China's oil and trade possibilities," Glasgow, 26 June 1975), p. 17.

26. For specifications for the Scarabeo III and others in this series, see *Offshore Drilling Register* (London: H. Clarkson, 1975), p. 66. See also "Three-caisson 'Scarabeo III' Joins the Offshore Fleet," *Ocean Industry* (September 1975), pp. 249–51.

27. *Petroleum News Southeast Asia* (1974), 5:2.

28. Matsuishi, "Chugoku no Engan Kaiyo Kaihatsu" [China's coastal oceanographic exploration], p. 42.

29. Cox, who has sold oil-related computer equipment to the Chinese, has established that the type of Sercel equipment acquired from CGG was the "pulse-width analog" variety, which is adaptable for offshore use.

30. The 1972 transactions are recalled in J. A. Kiely, *People's Republic of China, Bureau of Geology, Mission to Canada, September–October 1975*, Department of Industry, Government of Canada (Ottawa, 1975), p. 2.

31. Peter Hood, *A Visit to China* (Ottawa: Geological Survey of Canada, 1974), p. 45.

32. Japanese sources said that Tokyo-based CGG negotiators concluded a contract for the complete packages of equipment for two more boats. However, this would be confirmed from CGG or other sources.

33. "Wo Kuo Hai-yang Ti-chih Tiao-Ch'a Shih-ye Hsün-su Fa-chan" [Our work of exploring the ocean's geology has developed rapidly].

34. Hood, *Visit to China*, pp. 45–46.

35. *Ibid.*, p. 46.

36. *Ibid.*, p. 20.

37. *Ibid.*, pp. 36–37.

38. Article 3571.

39. *U.S. China Business Review* (September/October 1975), especially pp. 5, 27.

40. "Earthquake Research in China," *Transactions,* American Geophysical Union (November 1975), p. 846.

41. "Wo Kuo Hai-yang Ti-chih Tiao-Ch'a Shih-ye Hsün-su Fa-chan" [Our work of exploring the ocean's geology has developed rapidly].

42. "Earthquake Research in China."

43. The computers were ordered through the French subsidiary of Control Data but were covered by a U.S. export license and were to be manufactured in the United States.

44. "End-use" agreements in cases of this nature can range from the presence of resident U.S. inspectors or periodic visits by inspectors to the actual transfer of the computer records. Variants of these formulas have been used in computer sales to Soviet bloc countries, but Peking resisted any inspection until late 1976, when agreement was reached on an undisclosed formula ruling out resident inspectors. See a discussion of the screening procedures in sensitive technology transfers in "C.D.C. Gets the Green Light," *U.S. China Business Review* (November/December 1976), p. 51.

45. For a description of Model 111, see F. E. Allen and J. T. Schwartz, "Computing in China: A Trip Report" (July 1973), pp. 17–18 and Appendix 1; and Thomas E. Cheatham, Jr., "Computing in China: A Travel Report," *Science,* 12 October 1973, p. 140.

46. Raphael Tsu, "High Technology in China," *Scientific American* (December 1972), pp. 13, 16.

47. New China News Agency (cited in *Asia Research Bulletin,* 31 May 1975, p. 84) said that the new system consisted of two principal computing machines, fourteen small affiliated computers, and input-output facilities, with a total of 80 integrators and 1,000 operational amplifiers.

48. Cheatham et al., "Computing in China," p. 140.

49. *Ibid.*, p. 137, discusses Chinese efforts to make emitter-coupled

logic (ECL) integrated circuits. For a more skeptical view of Chinese computer capabilities, see Wade B. Holland, "Perspectives on Chinese Computing," *Soviet Cybernetics Review* (January 1973), pp. 22–24.

50. This statement was related by Cecil Craft, president of Seisdata of Houston, who lectured to the visiting Chinese group in November 1973.

51. Jan-Olaf Willums, "China's Offshore Oil: Application of a Framework for Evaluating Oil and Gas Potentials under Uncertainty" (Ph.D. dissertation, Massachusetts Institute of Technology, 1975), p. 309.

52. This has been communicated by contacts in the Chinese oil industry to Paul H. Fan, professor of geology at the University of Houston.

53. Colina MacDougall, "China Expands Port Facilities," *Financial Times,* 13 January 1976, p. 4.

54. Bobby A. Williams, "The Petroleum Industry in China: A Note" (unpublished paper, March 1976), p. 32.

55. "Bokkai (Daiko) Kaitei Yuden no Kaihatsu Katei" [Offshore oil field development in the Po Hai Gulf (Takang)], p. 7. See also "China's Oil Output May Triple by 1980," *Journal of Commerce,* 17 March 1975, p. 9. "New Details Disclosed on China's Oil" (*Journal of Commerce,* 30 April 1975, p. 2) makes the unsubstantiated assertion that "30 or more" production wells were then in operation in the Po Hai Gulf.

56. *Nihon Kogyo* (Tokyo), 17 June 1975, p. 6.

57. Dresser Industries agreed to provide spare parts for five years. China limited the training program for its technicians to a six-week course in Houston and a thirty-day instruction period after delivery. After that, Peking would have to contract anew for additional instruction.

58. Exploration has centered to the north of the island to determine whether existing onshore structures in the Maoming area extend offshore but has also included areas to the south and east.

59. A Canton television film publicizing China's claims to the Paracels showed a drilling rig on Yung-hsing Island (Woody Island) (*Foreign Broadcast Information Service Daily Report,* 7 June 1974,

p. H-2), and U.S., Soviet, and British intelligence sources agree that offshore drilling is also under way in the Paracels area.

60. "Over 20 Million B/D in the 1980's?" *Petroleum Economist* (London, February 1976), p. 50, Table II.

61. Willums, "China's Offshore Oil," especially p. 158, Table 5.6, and pp. 181–93, 257–59, 437–39.

62. For a detailed review of the Aga experience, see *Nihon Kaiyo Sekiyu Shigen Kaihatsu* [Japan's development of offshore oil resources], Tairikudana Sekiyu Tennen Gasu Shigen Kaihatsu Kondankai [Conference for the development of oil and natural gas resources on the continental shelf] (Tokyo, November 1975), especially pp. 4, 34–35, 93–94.

63. Hood, *Visit to China*, pp. 42, 45. Hood states that this statement was made in a meeting on 6 May 1974 by officials of the Technical Import Corporation. Hood told the author that the Chinese also expressed their interest in exploring in waters up to 900 feet in depth.

5. Offshore Oil
and the Future of Taiwan

1. Long known as ECAFE, the Economic Committee for Asia and the Far East was renamed ESCAP (Economic and Social Commission for Asia and the Pacific) in 1975.

2. Li studied geology on the mainland with many of the Chinese geologists who later fled to Taiwan and became the leading officials of the Chinese Petroleum Corporation there. C. Y. Meng, now CPC's chief geologist, was a member of Li's 1934 graduating class at Tsinghua University, where both studied geology, and the two later worked together at the Geological Survey of China. In 1944 Li was assigned as a Nationalist Chinese representative in the postwar reconstruction agencies then emerging. Five years later he joined ECAFE at its inception and was serving as the deputy director of its Division of Industry and Mineral Resources when he launched the CCOP idea in 1965.

3. "Inaugural Address by His Excellency Mr. K. T. Li, Minister for Economic Affairs, Republic of China," *Report of the Fourth Session of*

the CCOP, United Nations ECAFE (E/CN.11/L.190), 31 January 1968, Appendix III, p. 25. See also the speech by Vice President and Premier C. K. Yen, Appendix II, p. 23.

4. A Gulf representative later intercepted him when the *R. V. Hunt* reached Japan, Emery recalled, and asked to see his survey results. Emery said that he politely declined on the grounds that the CCOP had pledged to keep the findings confidential until the survey report was publicly released.

5. The text of the formal announcement describing in detail the boundaries of the claimed areas may be found in "Undersea Oil Hunt in Progress," *China Post* (Taipei), 16 October 1970, p. 4. These boundaries were shown in a U.S. State Department map (no. 154) issued by the Office of the Geographer in January 1971. By July 1971 the State Department had issued a new map (no. 261) corresponding to the more limited areas covered in Taipei's concession agreements. However, as late as 1975 Taipei continued to assert the validity of its original boundaries. See the letter from Y. F. Yang, vice-president for exploration, Chinese Petroleum Corporation, Taipei, in the widely read oil trade publication *Offshore* (March 1975), p. 19.

6. Chinese Petroleum Corporation officials state that their western concession boundaries are based solely on geological criteria and that Taipei retains the right to develop offshore resources up to the mainland coast when it has completed higher priority work.

7. In 1974 Amoco obtained a 50 percent interest in a $100-million joint venture with the Chinese Petroleum Corporation to manufacture terephthalic acid, used in producing polyester fibers.

8. Northcutt Ely, "Seabed Boundaries in the East China Sea" (advisory opinion submitted to Gulf Oil Company, 2 November 1970), p. 2.

9. *Ibid.,* p. 5.

10. In his 1970 opinion for Gulf, Ely, who also advises Taiwan on Law of the Sea issues, declared that the Senkakus (Tiao-yü T'ai) were "too insignificant" to be credibly used by Japan as base points in demarcating sea boundaries and that "on balance," Taiwan would probably be able to uphold its claims to areas up to the Okinawa Trough under then existing international law. Within a year, however, Ely had reversed himself in another opinion, "East

China Sea—Seabed Boundaries of Japan" (27 August 1971), especially pp. 1–18, 47–81, 103–5.

11. For details of these discoveries and subsequent drilling, see "Conoco Hits Gas and Condensate off Taiwan," *Oil and Gas Journal*, 2 September 1974, p. 28; "Discovery of Gas Well off Kaohsiung," *Asia Research Bulletin*, 31 December 1975, p. 157; and "Taiwan to Exploit Offshore Gas Field," *New York Times*, 2 June 1976, p. 42.

12. The three factors responsible for these estimates are the location of deposits in deep water, the scattered distribution of these deposits in pockets rather than big reservoirs, and the fact that the best pipeline route to shore might require spanning an undersea canyon. Kelly Brownlow, then president of the Amoco-Taiwan Petroleum Company, said in 1974 that recoverable reserves would have to approach 8 to 10 trillion cubic feet in order to merit development, with a precise estimate linked to the price to be paid by Taiwan.

13. Liquefaction for export would be unprofitable, it was felt, given the high development costs.

14. The Overseas Private Investment Corporation (OPIC) rejected Conoco's overtures for an investment guarantee. Although generally unwilling to guarantee offshore operations beyond a distance of twelve miles, OPIC noted that the waters concerned are disputed, which in itself ruled out U.S. government involvement in this instance.

15. H. T. Chiu, "Development of the Neogene Sedimentary Basin and Formation of Oil and Gas Fields in Northwestern Taiwan," *Petroleum Geology of Taiwan*, no. 10, Taiwan Petroleum Exploration Division, Chinese Petroleum Corporation, Chinese Petroleum Institute (December 1972), p. 36.

16. For example, see C. Y. Meng, "A Conception of the Evolution of the Island of Taiwan and Its Bearing on the Development of the Neogene Sedimentary Basins on Its Western Side," *CCOP Technical Bulletin*, United Nations ECAFE (Bangkok, May 1970), 3:109. See also C. Y. Meng, "Geologic Concepts Relating to the Petroleum Prospects of the Taiwan Strait," *CCOP Technical Bulletin*, United Nations ECAFE (Bangkok, June 1968), 1:145; and C. Y. Meng and J. T. Chou, "The Petroliferous Taiwan Basins in the Framework of the Western Pacific Ocean" (paper submitted by the Chinese Petro-

leum Corporation to the Circum-Pacific Energy and Minerals Resources Conference, Honolulu, 26–30 August 1974), pp. 5, 9. K. O. Emery has also shown that a long and nearly continuous ridge extends between Taiwan and the mainland in a direction roughly congruent with the Luichow Basin. See K. O. Emery and Zvi ben-Avraham, "Structure and Stratigraphy of the China Basin," *CCOP Technical Bulletin*, United Nations ECAFE (Bangkok, July 1972), 6:136.

17. The investor obtains an immediate deduction up to the amount of his investment for intangible drilling and exploration costs. If no producing wells result, his loss is limited to the after-tax cost. If there are successful wells, he can sell his interest at a capital gains rate, converting his ordinary deduction to a capital gain.

18. For a discussion of these legal battles, see Stanley Penn, "Mr. Anderson's Deals," *Wall Street Journal*, 3 April 1973, p. 38.

19. Recounting the charges by the deputy chief of staff of the Chinese air force office in Washington, General P. T. Mow, that some $7 million had been misappropriated by corrupt officials in Taipei, *The Reporter* pointed to J. Z. Huang as the trusted emissary sent by Chiang Kai-shek to offer a $300,000 rakeoff to silence General Mow. See "Cast of Characters: The China Lobby—Part II," *Reporter*, 29 April 1952, p. 4. See also "The Ubiquitous Major: The China Lobby—Part I," *Reporter*, 15 April 1952, p. 24.

20. "Registration Statement under the Securities Act of 1933," Amendment no. 1 to Form S-1 (statement submitted by Clinton International Corporation to the Securities and Exchange Commission, Washington, D.C., 8 March 1971). The statement declared that "for services rendered in securing rights for the company under the Taiwan Agreement, an unaffiliated individual has received a fee and reimbursible expenses aggregating $85,300 . . ." (p. 15). The list of exhibits at the end of the statement identified J. Z. Huang as the recipient of the fee (p. II-4), and Exhibit 13-F contained the text of a letter from Vice President P. E. Baria declaring that "the successful negotiations covering this agreement were, to a great extent, due to your efforts." Baria to Huang (% David Huang, 53 Nanking East Road, Taipei), 16 August 1970.

21. Setting forth the terms of the agreement with Taiwan, the statement explained that the concession would cover 13,400 square miles

initially but would be reduced to 5,760 square miles following detailed seismic surveys scheduled to end in March 1972. The company was committed to spend $4 million in phased-work obligations by March 1975 and would then be committed to spend an additional $7 million for further exploratory work. At the end of eight years the company would relinquish its entire holdings if commercial production had not been initiated. These terms were similar to those in other concession agreements with Taipei.

22. Exhibit 13-F contains the texts of the letter from Baria to Huang, cited earlier, and of a royalty agreement with Huang dated 26 September 1970.

23. At the time Findeiss met with McConaughy, Clinton's seismic contractor was forced to use SHORAN (short-range) navigation equipment based on nearby fishing junks, a method considerably less accurate than satellite navigation, in accordance with the overall U.S. effort to discourage exploration in disputed waters (see chapter 1).

24. In January 1973 the Securities and Exchange Commission filed a complaint against the company, charging numerous securities law violations and suspending over-the-counter trading for a year. The SEC charged "an elaborate scheme to improve the appearance of the company's financial statements, bolster the price of its stock, and enrich company insiders at the expense of public shareholders." (See "SEC Sues Clinton Oil," *Wall Street Journal*, 16 January 1973, p. 4.) Nine shareholder class-action lawsuits were also filed against the company and were not settled until June 1975. (See "Clinton Oil Settlement of Holder Class Suits Is Approved by Judge," *Wall Street Journal*, 2 June 1975, p. 23.)

25. The agreement is noted in "International Briefs," *Oil and Gas Journal*, 13 May 1974, p. 34.

26. Owned by Attwood Oceanics, the *Margie* had been leased to Atlantic-Richfield for use off Australia prior to the sublease arrangement with Shell, but Atlantic-Richfield no longer needed the rig and did not want to renew its lease in the absence of a long-term sublease arrangement.

27. Superior had an option to drill within a specified period, officials said, but had not made an ironbound commitment to do so, and in any case the agreement between Superior and Clinton had

not been formally approved by the Chinese Petroleum Corporation as a result of disputes over its tax provisions.

28. This attitude is reflected in "The President's Message," *Annual Report* (1973), Oceanic Exploration Company, p. 3.

29. For a sketch of Chiang's career and association with Oceanic, see "Oceanic Exploration Company (Taiwan)," *Beautiful China* (Taipei, July 1974), p. 25.

30. Registration No. 2-41051, 15 October 1971, p. 42. This median line and a subsequent one suggested by Oceanic utilize base points different from Ely's line for Gulf and would give Taiwan more extensive offshore jurisdiction.

31. This supplement appeared on 28 April 1972.

32. "Oceanic Exploration Company" (research report prepared by J. D. Mote, Research Division, Bache and Company, 20 June 1973), pp. 5–6.

33. *Annual Report* (1973), Oceanic Exploration Company, pp. 3, 11.

34. Oceanic had itself become increasingly chary of the political problems associated with the Taiwan concession and had relinquished the northern and western portions of its own concession area involving the greatest danger of conflict with Japan and China, respectively. Offshore oil agreements normally provide for the periodic relinquishment of territory following specified exploration phases.

35. *1974 Interim Report*, Oceanic Exploration Company, 21 November 1974.

36. In order to obtain a $7-million Thailand concession, Grynberg had borrowed $5.5 million from Texasgulf and was able to repay this debt, after a series of much-publicized postponements, only by borrowing anew outside the United States. Oceanic's hopes were fired by a series of strikes in the Aegean Sea off the coast of Greece, first a natural gas strike in May 1973, and then a major oil discovery at Prinou in February 1974. But the company was forced to renegotiate the terms of its concession as a result of political changes in Greece and the rise in crude oil prices after 1973, complicating its efforts to obtain more than $200 million in financing needed to get the Greek concessions into actual production. In 1976 it sold out its

Greek holdings entirely. See "Oceanic Concession Revised," *Petroleum Economist* (London, April 1975), p. 148; and *Ocean Oil Weekly Report*, 6 September 1976.

37. So named for its owner, David B. "Tex" Feldman.

38. The company delayed action on the provisional agreement reached in August 1971, until finally signing a formal agreement on 17 June 1972. The concession took legal effect on 29 August 1972.

39. "A Fine Edwardian Fling in Filmland," *Life*, 14 January 1957, pp. 126–27.

40. Comoro, General Crude, Weeks Natural Resources, Canadian Superior, Western Decalta (also Canadian), and Bochumer Mineral of West Germany.

41. *Asia Research Bulletin*, 31 January 1976, pp. 165–66.

42. P. D. Gaffney, C. P. Moyes, and B. Aling (Gaffney, Cline and Associates), "Economic Appraisal of the Potential Petroleum Resources of the Asian Pacific Region" (paper delivered at the Offshore Southeast Asia Conference, Singapore, February 1976), p. 26.

43. One of these resulted in a discovery of undisclosed size in fifty feet of water near Lukang, off central Taiwan, in December 1975.

44. A. A. Addington, "Taiwan," *Ocean Industry* (April 1975), pp. 143, 145. Y. F. Yang, vice-president for exploration of the CPC, said that production platforms would be given priority over rigs (Yang to Harrison, 9 February 1976).

45. Clinton, Conoco, and Amoco expire, respectively, in March, July, and September of 1978. Oceanic expires in March 1979, Gulf in March 1980, and Texfel in August 1980.

6. Offshore Oil
and the Future of Korea

1. For a summary of South Korean energy projections by Commerce Minister Chang Ye-chun, see "South Korea's Energy Development Program (1974–81)," *Asia Research Bulletin*, 31 July 1974, pp. 68–69.

2. A South Korean presidential proclamation on 18 January 1952 defining Seoul's claims over the offshore seabed is recorded in

United Nations, Secretariat, *Laws and Regulations on the Regime of the Territorial Sea* (ST/LEG/SER.B/6), November 1957, p. 30.

3. For examples of early CCOP projects in South Korea, see "Results of Exploration Relating to Offshore Prospects in the Republic of Korea" (pp. 57–61) and "Offshore Exploration Projects in the Republic of Korea" (pp. 80–82), *Report of the Fourth Session of the CCOP*, United Nations ECAFE (E/CN.11/L.190), 31 January 1968.

4. *The Petroleum Industry and Korea, 1968* (Seoul: Korea Oil Corporation, December 1968), p. 26. See also "Map Showing Postulated Sea Bottom of Korean Continental Shelf," p. 27, Fig. 8-1.

5. K. O. Emery et al., "Geological Structure and Some Water Characteristics of the East China Sea and the Yellow Sea," *CCOP Technical Bulletin*, United Nations ECAFE (Bangkok, May 1969), 2:4.

6. *Petroleum Industry and Korea, 1968*, pp. 28–29.

7. John J. McCloy discusses the origins of Gulf's involvement in Seoul in his report to the board of directors of Gulf concerning $4 million in illegal political contributions to Park's ruling party. *Report of the Special Review Committee of the Board of Directors of Gulf Oil Corporation*, U.S. District Court for the District of Columbia, 30 December 1975, pp. 96–98.

8. Article 3 of the Submarine Mineral Resources Development Law, enacted on 1 January 1970, authorizes the designation of concession boundaries by presidential decree. The coordinates for the sea boundaries set forth in the Presidential Enforcement Decree of 30 May 1970 may be found in Shigeru Oda, "The Delimitation of the Continental Shelf in Southeast Asia and the Far East," *Ocean Management* (December 1973), 1:346.

9. For definitive discussions of the South Korean approach to the demarcation of its boundaries, see Choon-ho Park, "Oil Beneath Troubled Waters," *Harvard International Law Journal* (Spring 1973), 14:236–48; and "The Sino–Japanese–Korean Sea Resources Controversy and the Hypothesis of a 200-Mile Economic Zone," *Harvard International Law Journal* (Winter 1975), 16: especially pp. 35–38.

10. Chong-su Kim (Geological and Mineral Institute of Korea), "The Petroleum Potential of the Korean Offshore" (paper delivered

at the Circum-Pacific Energy and Minerals Resources Conference, Honolulu, 26–30 August 1974), p. 6.

11. S. B. Frazier et al., "Marine Petroleum Exploration of the Huksan Platform: Korea" (paper delivered at the Circum-Pacific Energy and Minerals Resources Conference, Honolulu, 26–30 August 1974), pp. 3, 13–14 and Figs. 1–4.

12. *Ibid.* See also Kim, "Petroleum Potential of the Korean Offshore," especially pp. 9–10 and Fig. 4; and Technical Advisory Group (CCOP), "Geologic Basis for Locating Seismic Refraction Profiles in Offshore Areas of the Republic of Korea, Japan, Republic of China and the Philippines," Agenda Item 3, *Report of the Second Session of the CCOP,* UN ECAFE (E/CN.11/L.168), 10 December 1966, pp. 57–58.

13. Part of this concession area fell within the scope of the joint South Korean–Japanese agreement concluded in 1974. Exploration in the joint development area was suspended pending ratification of the agreement by the legislatures of the two countries, and this suspended status gave Texaco a legal justification for its failure to drill in Zone IV.

14. American Embassy (Seoul) to Secretary of State, Department of State Telegram 120850Z, Confidential 668, 12 April 1974. Made available in response to inquiries under the Freedom of Information Act.

15. The islands in question are discussed in Donald R. Allen and Patrick H. Mitchell, "The Legal Status of the Continental Shelf of the East China Sea," *Oregon Law Review* (Summer 1972), 51:796–802 (see in particular Table 1, p. 800). Allen and Mitchell were both working at the time this article was written for Northcutt Ely, Gulf's Law of the Sea adviser.

16. Hodgson to Harrison, 17 November 1975.

17. Northcutt Ely, "Boundaries of Seabed Jurisdiction off the Pacific Coast of Asia" (paper presented at the Circum-Pacific Energy and Minerals Resources Conference, Honolulu, 26–30 August 1974), p. 13.

18. *Ibid.,* p. 19.

19. In the company's *Interim Report to Stockholders,* no. 2 (9 June 1975, p. 3), Harrison reported that Zapata's drilling program in

South Korea had been "suspended indefinitely because of unresolved boundary disputes and other problems beyond the control of the company."

20. According to South Korean sources, Gulf had completed $6.5 million of a required $8 million work obligation by the end of 1974. Gulf states that the gap was smaller but declines to specify a figure. Under the original timetable in its 1967 agreement, Gulf was required to fulfill its contractual obligations by April 1977.

21. *Journal of Commerce*, 27 May 1975, p. 10.

22. "Geology of ROK Coast Favorable for Oil Find: Expert View," *Korea Herald* (Seoul), 8 January 1976, p. 1.

23. For a discussion of South Korean and Japanese approaches to Law of the Sea issues, see Park, "Oil Beneath Troubled Waters," pp. 238–48.

24. Korean Central News Agency (Pyongyang), 3 February 1974.

25. *Peking Review*, 8 February 1974, p. 3.

26. *Mainichi* discusses the Chinese claims in "Problematical Points in Japan-ROK Shelf Agreement," 16 March 1975, p. 4.

27. "Large Oil Layer in the East China Sea: Japan–Korea Agreement Is the Key," *Kagaku Kogyo Nippo* [Chemical industry daily], 26 September 1974, p. 1.

28. For examples of this argument, see "Government Desirous of 'Continuation of Deliberations' on Japan–ROK Shelf Agreement," *Mainichi*, 27 June 1975, p. 1; and "Do Not Ratify Continental Shelf Agreement in Haste," *Yomiuri*, 15 March 1975, p. 4.

29. "Problematical Points in Japan–ROK Shelf Agreement."

30. "Japan–ROK Agreement and National Interests," *Asahi*, 7 April 1975, p. 4.

31. "Joint Oil Development in East China Sea: ROK to Develop It 'Even by Itself,' " *Yomiuri*, 14 September 1975, p. 9.

32. "Government Desirous of 'Continuation of Deliberations.' "

33. Joseph S. Chung, *The North Korean Economy: Structure and Development* (Stanford, Calif.: Hoover Institution Press, 1974), p. 107, Table 29.

34. "Korea Needs 'Very Subtle Diplomacy,' " *Korea Herald*, 19 December 1975, p. 2.

7. China, Japan,
and the Oil Weapon

1. These figures are for the fiscal year ending 1 March 1975. In February 1976 Minoru Masuda, director-general of the Energy Resources Agency, referred to a 59.1 percent dependence on the majors. See "Will Promote 'Separation from Majors': Energy Agency Director," *Sankei* (Tokyo), 11 February 1976, p. 9.

2. *Nihon Keizai* (Tokyo), 17 November 1974.

3. "Oil and Materials: Japan's Problems and Policies," *US/Japan Outlook* (Fall 1974), p. 7.

4. For a representative sample of the bitterness toward the majors marking discussions of oil policy in the Japanese media, see a special issue of the weekly *Ekonomisuto* (Tokyo), 10 February 1976. See especially Saburo Kugai, "Ajia Seiatsu o Hakaru Meija" [Major oil companies plan pressure in Asia], pp. 24–29.

5. I discuss this relationship in detail in a forthcoming study, *The Widening Gulf: Asian Nationalism and American Policy* (New York: Free Press, 1977) and in a paper prepared for the 1975 Seminar Series of the Washington Center of Foreign Policy Research, " 'The State of the Arc': Japan, China and Korea." For a preliminary analysis, see "China and Japan: The New Partnership," *Washington Post,* 4 March 1973, p. B5.

6. For an analysis of Sino-Japanese trade tensions during this period by Hiroshi Tanimura, deputy director of the Mitsubishi Corporation China Division, see "Japan-China Trade at a Turning Point," *Kaigai Shijo,* Japan External Trade Organization (Tokyo, October 1975). This is presented in translated form in *JETRO China Newsletter,* 16 January 1976, pp. 8–19. Note especially the discussion of the competition for the Japanese textile market between China and South Korea and the probable mitigating effect of expanded Chinese oil exports to Japan on this competition.

7. For a discussion of Chinese efforts to adapt to the Japanese market, see Alistair Wrightman, "How China Is Adapting to the Japanese Market," *U.S. China Business Review* (July/August 1974), pp. 30–33.

8. Keiji Samejima, "Chinese Crude Oil: Imports on the Way to a Large Increase," *Nihon Keizai* (Tokyo), 15 November 1975, p. 1.

9. Samejima, "Chinese Crude Oil," Table 1. Another study indicated a similar pattern between residual heavy oil and other distillates but showed a gasoline yield of only 10.1 percent in the case of Taching in contrast to 25 percent in the case of the Arabian crude. See the statistics provided by the Petroleum Producers Association of Japan in "Snags Facing China's Oil Exports," *Far Eastern Economic Review* (Hong Kong), 28 November 1975, p. 47.

10. "Business Talks on Chinese Oil Exports to U.S.," *Asahi* (Tokyo), 25 December 1975, p. 2.

11. See especially Kugai, "Ajia Seiatsu o Hakaru Meija" [Major oil companies plan pressure in Asia], p. 25.

12. In "Ajia ni Utsutta Sekiyu Senso" [Oil war moves to Asia], a roundtable discussion in *Ekonomisuto* (Tokyo), 10 February 1976, p. 10.

13. Samejima, "Chinese Crude Oil," p. 2.

14. "Ajia ni Utsutta Sekiyu Senso" [Oil war moves to Asia], p. 14.

15. Cited in Samejima, "Chinese Crude Oil," p. 3.

16. "Japan: The Struggle Towards Recovery," *Petroleum Economist* (London, January 1976), p. 5.

17. See " 'Chinese Crude Oil' Heading Toward Long-Term Contract," *Yomiuri* (Tokyo), 20 September 1975, p. 1; and "Long-Term Imports of Chinese Crude Oil," *Yomiuri*, 24 December 1975, p. 9.

18. Memorandum, Masao Sakisaka (director, Institute of Energy Economics, Tokyo) to Harrison, 28 April 1975, Table 2.

19. "Japan and Chinese Crude Oil," *Japan Petroleum Weekly* (Tokyo), 4 August 1975, p. 3.

20. "Ajia ni Utsutta Sekiyu Senso" [Oil war moves to Asia], p. 15.

21. "Task Is How to Correct Imbalance in Japan-China Trade: Round Table Discussion by Keidanren Mission to China," *Nihon Keizai* (Tokyo), 29 October 1975, p. 9.

22. *Ibid.*

23. "For Balanced Expansion of Japan-China Trade," *Nihon Keizai* (Tokyo), 9 November 1975, editorial.

24. Geologist Michihei Hoshino stressed the likelihood of gas rather than oil in "The Continental Shelf of the East China Sea," *Okinawa*, Nampo Koho Engokai (Tokyo, December 1972), pp.

100–4. This Japanese monthly, now defunct, was published from 1968 to 1973. See also Yasufumi Ishiwada, "Nippon Retto Shuhen Tairiku Dankyu no Sekiyu Kishitsu" [Petroleum geology of the continental shelf surrounding the Japanese islands], *Sekiyu Gakkai-shi* [Oil Association journal] (Tokyo, 1975), 18:27.

25. Hiroshi Kono, "Higashi Shina-kai Jiriki Kaihatsu ni Chugoku no Hisometa Omowaku" [The hidden intentions behind China's plans to develop the East China Sea alone], *Zaikai Tembo* (Tokyo, August 1974), pp. 15–16.

26. "Japan's Energy White Paper," *Japan Petroleum Weekly* (Tokyo), 6 May 1974, p. 18.

27. Offshore-related budget provisions are detailed and analyzed in *Nihon Kaiyo Sekiyu Shigen Kaihatsu* [Japan's development of offshore oil resources] Tairikudana Sekiyu Tennen Gasu Shigen Kaihatsu Kondankai [Conference for the development of oil and natural gas resources on the continental shelf] (Tokyo, November 1975), p. 56.

28. *Ibid.*, p. 5.

29. *Ibid.*, p. 26.

30. *Ibid.*, p. 56.

31. For example, see Y. Otsuka, "Tertiary Crustal Deformations in Japan," *Jubilee Publication on the Commemoration of Professor H. Yabe's Sixtieth Birthday* (Tokyo: Nagata Press, 1939), 1:481–519.

32. The author was resident in Tokyo during this period. See also an allusion to the impact of oil prospects on the Japanese Okinawa demand in Tao Cheng, "The Sino-Japanese Dispute Over the Tiao-Yü-Tai [Senkaku] Islands and the Law of Territorial Acquisition," *Virginia Journal of International Law* (Winter 1974), 14:242.

33. K. O. Emery and Hiroshi Niino, "Stratigraphy and Petroleum Prospects of Korea Strait and the East China Sea," *Report of Geophysical Exploration* (Seoul: Geological Survey of Korea, June 1967), pp. 1, 5.

34. See Daisuke Takaoka, "Participating in a Scientific Survey of Waters around the Senkaku Islands," *Okinawa*, Nampo Doho Engokai (Tokyo, March 1971), pp. 42–65.

35. "Marine Geologic Investigations of the Offshore Area around the Senkaku Islands," Technical Advisory Group, Agenda Item

4(b), *Report of the Seventh Session of the CCOP,* United Nations ECAFE (E/CN.11/L.278), 23 July 1970, pp. 99, 111. See also "Japan's Oil Development Abroad Has Fatally Fallen Behind Others," *Diamond* (Tokyo), 5 October 1970.

36. "A Roundtable: Senkaku Islands Oil Resources Awaiting Development," *Okinawa,* Nampo Doho Engokai (Tokyo, March 1971), pp. 25–40.

37. This plan is spelled out in "Japan Worried Over Peking View of Joint Oil Link with Taiwanese," *Yomiuri Daily News* (Tokyo), 3 November 1970, p. 1.

38. Teikoku is specifically cited in "Joint Oil Venture Hinted in Pact on Senkakus," United Press (Taipei), 21 October 1970 (*Japan Times,* 21 October 1970). This was confirmed by the author in Tokyo at the time. For other accounts underlining the inclusion of the Senkakus (Tiao-yü T'ai) in the scope of proposed joint development plans, see "Japan, Korea, Taiwan Okay Oil Project," *Asahi Evening News* (Tokyo), 16 November 1970, p. 1; and "Ocean Development by Japan, R.O.K. and Taiwan Decided by Liaison Committee: East China Sea, Including Senkaku Islands," *Nihon Keizai* (Tokyo), 22 December 1970, p. 1.

39. *Peking Review,* 11 December 1970, p. 15.

40. See *Nihon Kaiyo Sekiyu Shigen Kaihatsu* [Japan's development of offshore oil resources], especially p. 28 and Table IV, pp. 32–33.

41. Tsunenobu Omija, "Okinawa Kaiten Dai-yuden wa Waga te no Naka ni ari" [Okinawa's big undersea oil fields are in my hands], *Gendai* (Tokyo, March 1975), pp. 358–63. See also "Uruma Resources Starts Prospecting Islands Off Okinawa" in the business weekly *Nikkan Kogyo* (Tokyo), 1 July 1974, p. 1.

42. See chapter 9, p. 223. The natural-prolongation principle, as enunciated by the International Court of Justice in the North Sea cases, is discussed extensively in Choon-ho Park, "Oil Beneath Troubled Waters," *Harvard International Law Journal* (1973): especially 14:235–36.

43. Ishiwada, "Nippon Retto Shuhen Tairiku Dankyo no Sekiyu Kishitsu" [Petroleum geology of the continental shelf surrounding the Japanese islands], p. 26. See also *Nihon Kaiyo Sekiyu Shigen Kaihatsu* [Japan's development of offshore oil resources], pp. 23, 27.

44. The Japanese Foreign Ministry formally spelled out its stand on 8 March 1972 in a mimeographed statement, "The Foreign Ministry's View Concerning the Rights to Ownership Over the Senkaku Islands." For a sample of the voluminous literature presenting the Japanese approach to the Senkaku Islands (Tiao-yü T'ai) issue, see Toshio Okuhara, "The Territorial Sovereignty Over the Senkaku Islands and Problems on the Surrounding Continental Shelf," *Japanese Annual of International Law* (1971), 15:97–105.

45. The United States has declared that the islands were captured as a prize of war and that returning them to Japan did not constitute recognition of Japanese, as against Chinese, territorial claims. See the discussion of the Senkaku (Tiao-yü T'ai) issue in *Okinawa Reversion Treaty*, Hearings Before the Committee on Foreign Relations, U.S. Senate, 27–29 October 1971, especially pp. 88–93, 144–54.

46. The Chinese position on the Senkaku Islands (Tiao-yü T'ai) is analyzed in Cheng, "Sino-Japanese Dispute Over the Tiao-yü T'ai Islands," especially pp. 248–60. See also Victor H. Li (Stanford University Law School), "China and Off-shore Oil: The Tiao-yü T'ai Dispute" (paper prepared in August 1974). This paper has been published in abridged form under the same title in the *Stanford Journal of International Studies* (Spring 1975), pp. 143–58.

47. This contrast is underlined by Tadao Ishikawa in "Should Not Mix Up Northern Territory Problem and Senkaku Problem," *Sankei* (Tokyo), 19 November 1974, p. 7.

48. A copy of Ushiba's letter, dated 3 December 1971, is in my possession.

49. A copy of Clinton's position paper for the Findeiss-McConaughy meeting is in my possession.

50. Kazuo Yatsugi, "Nikkan Choki Keizai Kyoryoku Shian" [Long-range economic cooperation between Japan and Korea] and "Nikkan Goben Narabi ni Kaku Boeki Shinko Koshi Setsuritsu-an" [Tentative plan for the establishment of a Japan-Korea joint venture and processing trade promotion corporation] (April 1970, privately circulated. Submitted to a meeting of the Japan-Korea Cooperation Committee, Seoul, 10 April 1970). As a result of Korean information leaks to *Dong-a-Ilbo* and other Seoul newspapers, a sharp Korean nationalist reaction forced withdrawal of the plan in its original form, but many of its proposals have been partially adopted in

piecemeal form. For an earlier account by the author, see "Japan's Yen Buys Little Love in Korea," *Washington Post,* 27 February 1973, pp. A1, A12, one of a series of eight articles on Japanese influence in Asia.

51. Kazuo Yatsugi, "Kaiyo no Kyodo Kaihatsu An" [A plan for joint ocean development] (21 December 1970, privately circulated).

8. Danger Zones
in the South China Sea

1. Shih Ti-tsu, "South China Sea Islands, Chinese Territory since Ancient Times," *Peking Review,* 12 December 1975, p. 10. This article was reprinted from *Kuangming Jih-pao* (24 November 1975) with the addition of a map indicating the extent of Chinese boundary claims in the South China Sea.

2. For example, see *Petroleum News Southeast Asia* (Singapore, February 1974), p. 32; "The South China Sea Islands," *China Reconstructs* (Peking, August 1974), pp. 19–22; and "Construction in the Hsishas," *China Pictorial* (Peking, October 1975), p. 5.

3. Shih, "South China Sea Islands," p. 10.

4. New China News Agency, 11 January 1974.

5. The geological environment in the South China Sea is described more fully in chapter 3. See also M. L. Parke, K. O. Emery et al., "Structural Framework of the Continental Margin in the South China Sea," *CCOP Technical Bulletin,* United Nations ECAFE (Bangkok, June 1971), 4:103–42.

6. Chinese historical claims are detailed with copious supporting citations from Chinese sources in Tao Cheng, "The Dispute over the South China Sea Islands" (paper made available to me). See also Hungdah Chiu and Choon-ho Park, "The Legal Status of the Paracel and Spratly Islands," *Ocean Development and International Law* (1975): especially 3:16–30.

7. Shih, "South China Sea Islands," p. 13.

8. The Vietnamese case as defined by the Thieu regime is most completely presented in *Hoang-sa* [The Paracel Islands] (Saigon: Vietnam Cong Hoa Bo Dan Van Va Chien Hoi, 20 March 1974). See

also Chiu and Park, "Legal Status of Paracel and Spratly Islands," especially p. 11–16.

9. Chiu and Park, "Legal Status of Paracel and Spratly Islands"; and Cheng, "Dispute over the South China Sea Islands," entirety.

10. The Malaysian reaction to the Tseng-mu claim and the *Red Star* salvo is reflected in "Renewed Chinese Claim on Tsengmu Surprises Kuala Lumpur: Tiny Sand Reef Makes Big Waves," *Hong Kong Standard,* 14 December 1974, p. 7. For useful summaries of Sino-Soviet exchanges over South China Sea claims, see Nayan Chanda, "Sino-Soviet Rivalry: Islands of Friction," *Far Eastern Economic Review* (Hong Kong), 12 December 1975, pp. 28–29; and Donald R. Bakke, " 'Historical Territorial Rights' Key to China's Future," *Offshore* (February 1976), pp. 76–78.

11. Chanda, "Sino-Soviet Rivalry," p. 29.

12. The Agence France-Presse report is summarized in "North Vietnam's Petroleum Agreement," *Petroleum News Southeast Asia* (Singapore, October 1974), p. 47.

13. Letters: S. Orioli (president, AGIP-Mineraria) to Harrison, 16 and 28 April 1975.

14. The semiofficial Central News Agency in Saigon, reporting the decree annexing eleven of the Spratlys, said on 24 September 1973 that the move had been prompted by a recommendation of the Petroleum and Mineral Agency.

15. *Petroleum in Vietnam,* Petroleum and Mineral Agency (Saigon, February 1975), p. 5.

16. For example, see an account of life in the Paracels in *China Pictorial* (Peking, October 1975), pp. 12–13.

17. See *China News Summary* no. 605, U.K. Regional Information Services (Hong Kong), 25 February 1976, p. 2.

18. This statement was issued in Saigon on 23 January 1974 by Major Phuong Nam, press spokesman for the PRG delegation to the Joint Military Commission.

19. United Press International (Bangkok), 23 October 1974.

20. *Nhan Dan* (Hanoi), 31 October 1974.

21. See the United Press International dispatch (Saigon), 10 No-

vember 1974 and a skeptical account of the incident in *Ocean Oil Weekly Report,* 18 November 1974, pp. 1–2.

22. "Saigon Acts to 'Sell Out' Natural Resources," Radio Hanoi (1115 GMT), 5 July 1973.

23. "The Inviolable Property of the South Vietnamese People," Radio Hanoi (1400 GMT), 21 February 1974.

24. "Foreign Ministry Denounces Their Oil Contracts," Liberation Radio (0302 GMT), 3 June 1974.

25. "Thieu's Oil Ballyhoo," Liberation Radio (2200 GMT), 30 August 1974.

26. Vietnam Thong Tan Xa [Vietnam press agency] (Saigon), 2 November 1974.

27. "The Transition of the Vietnamese Economy," reprinted in *Vietnam Press* no. 6864, U.S. Embassy (Saigon), 17 November 1974, p. B-3.

28. Philip A. McCombs, "Oil Find Cheers South Vietnam," *Washington Post,* 27 October 1974, p. A17.

29. *Petroleum in Vietnam,* p. 8.

30. Martin Woolacott, "Oil to the Rescue?" *Guardian,* 14 September 1974, p. 5.

9. East Asia in the
Law of the Sea Debate

1. See especially Choon-ho Park, "Oil Beneath Troubled Waters," *Harvard International Law Journal* (1973), 14:212–60; and Donald R. Allen and Patrick H. Mitchell, "The Legal Status of the Continental Shelf of the East China Sea," *Oregon Law Review* (Summer 1972), 51:789–812.

2. *Convention on the Continental Shelf,* Geneva, 29 April 1958, 1 U.S.T. 471, T.I.A.S. No. 5578, 499 U.N.T.S. 311 (1964). See especially Articles 1, 3, and 6.

3. *North Sea Continental Shelf Cases* (1969), International Court of Justice, 53.

4. *Peking Review,* 11 December 1970, pp. 15–16.

5. *Peking Review,* 27 November 1970, p. 7.

6. *Peking Review,* 11 March 1972, p. 14.

7. "Working Paper on Sea Area Within the Limits of National Juris-diction" (submitted by the Chinese delegation to Subcommittee II of the Committee on the Peaceful Uses of the Seabed and Ocean Floor Beyond the Limits of National Jurisdiction), U.N. Doc. A/AC.138/SCII/1.34, pp. 2–4.

8. Statement by Chinese delegate Huang Ming-ta before the CCOP during a discussion of Agenda Item 7(b)(i), April 5, 1974, at the thirtieth meeting of the United Nations Economic and Social Com-mission for Asia and the Far East (ECAFE), Colombo (Sri Lanka). A copy of the transcript was provided for me by CCOP. Note: ECAFE was renamed the Economic and Social Commission for Asia and the Pacific (ESCAP) in 1975.

9. New China News Agency, Release No. 49, New York, 13 June 1977.

10. Park, "Oil Beneath Troubled Waters," especially p. 240.

11. Northcutt Ely, "Boundaries of Seabed Jurisdiction off the Pacific Coast of Asia" (paper delivered at the Circum-Pacific Energy and Minerals Resources Conference, Honolulu, 26–30 August 1974), pp. 20–25.

12. For example, see "Joint Oil Development in the East China Sea: Government Racking Its Brains," *Yomiuri* (Tokyo), 14 September 1975, p. 9; and Susumu Awanohara, "Miki Faces a Seabed Wrangle," *Far Eastern Economic Review* (Hong Kong), 18 April 1975, p. 38.

13. Statement by Choon-ho Park in a discussion of "Seabed Dis-putes in East Asia" at the Association of Asian Studies Convention, San Francisco, 10 March 1975. This fear is based in part on a pointed Chinese U.N. reference to the "depth and inclination of the ocean floor" as a possible variable in determining offshore bounda-ries. United Nations, General Assembly, *Summary Record of the Fifty-fifth Meeting,* Subcommittee II, Committee on the Peaceful Uses of the Seabed and the Ocean Floor (A/AC/138/SCII), 20 March 1973, p. 4.

14. United Nations, Third Conference on the Law of the Sea, *Of-ficial Records,* 2d sess., Second Committee, 20th meeting, 30 July 1974, 1:162. "The boundary between states adjacent or opposite to

each other," said North Korean delegate Kim Guk-jun, "should be determined by consultation according to the principles of an equidistant or median line."

15. *International Legal Materials* (1971), 10:452.

16. United Nations, Third Conference on the Law of the Sea, *Revised Single Negotiating Text,* Part II, *Official Records* (May 1976), 5:164.

17. *Ibid.,* p. 72.

18. *Ibid.,* p. 164.

19. Hollis D. Hedberg, "Ocean Boundaries and Petroleum Resources," *Science,* 12 March 1976, pp. 1012–13. Hedberg, professor emeritus of geology at Princeton University, has served for many years as a regular adviser to Gulf.

20. In geological terms, Japanese geologists emphasize, the Okinawa Trough is a "back arc basin" related primarily to the Ryukyu Islands. See especially Sadanori Murauchi, "Rikyu Toko-kei no Chikyu Butsurigakuteki Kenkyu" [Geophysical studies on the Ryukyu Arc], *Kokuritsu Kahaku Senpo* [National scientific reports] (Tokyo), 20 September 1974, pp. 66–80.

21. A National Petroleum Council chart shows 70 percent of known offshore reserves to be in shelf areas, another 25 percent on the slope, and only 2 percent in the area between the base of the slope and the seaward edge of the rise. (See Richard Gardner, "Offshore Oil and the Law of the Seas," *New York Times,* 14 March 1976, Sec. 3, p. 4.) The chart fails to reflect proprietary information in the possession of individual oil companies concerning the potential of the rise. However, even the industry's public posture underlines the importance of the slope.

22. J. D. Moody, "An Estimate of the World's Recoverable Crude Oil Resources" (paper delivered at Panel Discussion 6(2), World Petroleum Congress, Tokyo, May 1975), p. 5.

23. Northcutt Ely, "Legal Problems in Undersea Mineral Development" (paper delivered at the American Institute of Mining, Metallurgical and Petroleum Engineers, Washington, D.C., 19 February 1969), p. 1.

24. *Petroleum Resources Under the Ocean Floor,* National Petroleum Council (Washington, D.C., March 1969), p. 10.

25. Northcutt Ely, "The American Consumer's Stake in the Law of the Sea," in Virginia S. Cameron, ed., *New Ideas, New Methods, New Developments,* vol. 2, *Exploration and Economics of the Petroleum Industry* (New York: Matthew Bender, 1974), p. 79.

26. Henry A. Kissinger first made this stand unambiguous in "International Law, World Order and Human Progress" (address before the American Bar Association Annual Convention, Montreal, 11 August 1975), p. 4. Informally, the United States has proposed that revenue-sharing begin in the sixth year of production at 1 percent of "wellhead market value," rising to 5 percent after ten years. This percentage is much lower than that proposed by most developing countries.

27. Hedberg, "Ocean Boundaries and Petroleum Resources," pp. 1011–12. Hedberg argues that revenue-sharing should not be based on oil production beyond a specified distance from shore.

28. Robert D. Hodgson, "National Maritime Limits: The Economic Zone and the Seabed," in Francis T. Christy, Jr. et al., eds., *Law of the Sea: Caracas and Beyond* (Cambridge, Mass.: Law of the Sea Institute, Ballinger, 1975), p. 190. Hodgson was referring to the original formulation of Hedberg's proposal in Hollis D. Hedberg, "National–International Jurisdictional Boundary on the Ocean Floor," *Occasional Paper No. 16,* (Kingston, R.I.: Law of the Sea Institute, University of Rhode Island, 1972).

29. See chapter 7, pp. 172–73.

10. The United States and Chinese Oil

1. The shifting U.S. policy during this period is noted in the *Peking Report* (report on a delegation to China from the National Council for U.S.-China Trade in November 1973), National Council for U.S.-China Trade Series, Special Report no. 6 (Washington, D.C., 12 December 1973), pp. 44–45. See also Nicholas Ludlow, "China, Oil and the United States, *U.S. China Business Review* (January/February 1974), p. 26.

2. *Peking Report,* p. 44.

3. *Ibid.,* p. 61.

4. *Ibid.,* p. 55.

5. *Ibid.*, p. 78.

6. *Petroleum Intelligence Weekly*, 16 June 1975, p. 10.

7. Statement by Vice Minister of Foreign Trade Yao I-lin in a meeting with former Japanese Foreign Minister Aiichiro Fujiyama, reported by the Kyodo (Japan) News Agency (Peking), 26 December 1975.

8. Charles K. Ebinger (Federal Energy Agency, Division of International Affairs), "Asian Energy Developments," mimeographed. Research study prepared for distribution primarily within the U.S. government and to members of Congress, Washington, D.C., 9 May 1975, p. 15.

9. Ronald Reagan, "Expanding Our Ties with China," *New York Times*, 28 July 1976, p. 31.

10. Eleanor B. Steinberg and Joseph A. Yager, *Energy and U.S. Foreign Policy* (Cambridge, Mass.: Ford Foundation Energy Policy Project, Ballinger, 1975), p. 222.

11. Paths to Peace and Development

1. This is an excerpt from a transcript of Chou's audience on 4 January 1973 with six overseas Chinese students in *Ch'i-shih Nien-tai* [The 1970s] (Hong Kong, April 1973), pp. 30–31. The transcript was signed by Yue Yu, believed to be a pseudonym for one of the students. This transcript was called to my attention by Choon-ho Park and was translated by Park and Richard Yiu.

2. *Peking Review*, 8 February 1973, p. 3.

3. Wang Yun-chi, "Tiao-yü T'ai Ch'ien-wan Tiao-pu-te" [Tens of thousands of "noes" to forsaking Tiao-yü T'ai], *Ming-pao Yüeh-k'an* [Ming Pao monthly] (Hong Kong, May 1971), pp. 27–29. The estimate of 80 billion barrels was attributed to an article in *Oil and Gas Journal*, 12 October 1970, p. 61.

4. Yung Ping, "Such Is the 'Chinese Petroleum Company,' " (Fukien broadcast to Taiwan in Mandarin, 1 November 1974), translated in *Foreign Broadcast Information Service Daily Report*, 4 November 1974, p. C1.

5. See especially *Ming-pao Yüeh-k'an* [Ming Pao monthly] (Hong Kong).

6. For a useful analysis of the "self-reliance" concept in relation to Peking's policy toward global resource issues, see Kim Woodard, "People's China and the World Energy Crisis," *Stanford Journal of International Studies* (Spring 1975), especially pp. 128–29. See also Wang, "Tiao-yü T'ai Ch'ien-wan Tiao-pu-te" [Tens of thousands of "noes" to forsaking Tiao-yü T'ai], pp. 23–24.

7. See Harned Pettus Hoose, "China Urges a Pacific Triumvirate," *Los Angeles Times*, 21 March 1976, Part IV, pp. 1, 4; and Ho Ping-ti, "China Is Now on the Way to Join the Oil Giants," *New Nation* (Singapore), 22 October 1974, p. 8.

8. This proposal is made by Donald R. Allen and Patrick H. Mitchell in "The Legal Status of the Continental Shelf of the East China Sea," *Oregon Law Review* (1972), 51:809. See also *International Legal Materials* (1969), 8:493–96; R. Young, "Equitable Solutions for Offshore Boundaries: The 1968 Saudi Arabia-Iran Agreement," *American Journal of International Law* (January 1970), 64:152–57; and *International Boundary Study, Iran-Saudi Arabia*, U.S. Department of State, Office of the Geographer, Series A., no. 24 (1970).

9. C. G. Jacobsen, "Japanese Security in a Changing World," *Pacific Community* (Tokyo, April 1975), pp. 361–62.

10. "Teng Hsiao-ping on the Situation in China and the Question of Taiwan—Notes Taken by Fan Lan and Others," *Ch'i-shih Nien-tai* [The 1970s] (Hong Kong, December 1974), pp. 15–17 (a transcript of Teng's conversation with a visiting group of overseas Chinese students).

11. See note 47, chapter 7.

12. Along with the United States, Britain and the Netherlands are also involved in South Korea, Japan, Brunei, and Malaysia through Royal Dutch Shell. Sweden and other West European countries are involved in Philippine concessions.

13. In 1975 ECAFE was renamed the Economic and Social Commission for Asia and the Pacific (ESCAP).

14. Initially, North Korea stands to gain the most from coordinated offshore efforts because survey results to date suggest that the most promising undersea reserves are located in areas adjacent to the

southern part of the peninsula. Many Korean offshore areas remain relatively unexplored, however, especially those adjacent to the North, and some geologists point to promising structural features off the North's coast in the Sea of Japan.

15. The 1971 agreement divided operational responsibility among the companies holding concessions that predated the accord in the areas involved and spelled out a complex formula for the allocation of costs and benefits. This formula offers possible guidelines for similar agreements in the future, although it would be necessary to take into account the pricing policies associated with governmental control of oil development in the cases of China, North Korea, and Vietnam.

Index